Fletcher, Task Force Commander:
The early months of the Pacific War
by James Bauer

Starting with Pearl Harbor, the Japanese were able to rampage throughout the Pacific where their Navy defeated five fleets of three nations in five months without losing a single ship. The book on aircraft carrier-on-carrier warfare had not yet been written when Admiral Frank Jack Fletcher made the first attack where opposing ships never saw each other. He was the man on the spot who had to invent that book.

After saving Australia in the Battle of the Coral Sea, he led two task forces at the crucial Battle of Midway where the overwhelming superiority of Japan's navy was reduced, and then he covered the invasions at Tulagi and Guadalcanal and saved that beachhead in the Battle of Eastern Solomons. In eight months of continuous combat, Fletcher sank six enemy carriers with the loss of only two in a near-perfect balance of aggressiveness and caution.

Fletcher deserves far more attention than he has been given in the public history of the Pacific War. This book helps spread the word while providing a rich reference to help understand that war.

"Fletcher, Task Force Commander" is in three parts. The first is about the early life of Fletcher raised in a small town in Iowa to an entrepreneurial family. The second part summaries the naval battles of the first year of the war that starts with the Japanese winning and has the appropriate title "Till they met Fletcher". The third part is a reference with hard to find lists, comparisons, specifications, photos, timelines and discussions on various topics about that first year of the war. Some myths of that period are exploded including some that are embedded in popular culture. As the author put it, "This is information one needs to understand the Pacific War".

Fletcher,
Task Force Commander:
the early years of WWII in the Pacific.

James L Bauer

Manorborn House
Marshalltown, Iowa

Fletcher, Task Force Commander:
the early years of WWII in the Pacific.

The story of
the dangerous times of
Admiral Frank Jack Fletcher.

The war in the Pacific was being lost to a rampaging enemy that had defeated five fleets of three nations until they came face to face with Fletcher leading the remnants of a depression era fleet.

James L. Bauer
Marshalltown, Iowa

Expanded Commemorative Edition

To those who served and gave their lives.

JAMES L. BAUER is a student of the early years of the Pacific War and a former sailor.

© Copyright 2010, 2018

        Manorborn Press,
        2490 248th Street,
        Marshalltown, Iowa, 50158

        10 9 8 7 6 5

First Edition, 2010   Commemorative

Second Edition, 2018 Expanded

Library of Congress Cataloging-in Publication Data

Bauer, James L.
Fletcher, Task Force Commander : the early years of WWII in the Pacific / James L. Bauer. – 2nd ed.

ISBN-10:   0-9830502-0-1
ISBN-13: 978-0-9830502-0-9

1. Fletcher, Frank Jack.
2. United States – History, Naval – 20th Century.
3. World War, 1939-1945 – Naval operations, American.
4. World War, 1939-1945 – Campaigns – Pacific Ocean

All rights reserved. The editorial arrangement, analysis, and commentary are subject to this copyright notice. No part of this book may be reproduced or transmitted in any form by any means, electronic or mechanical, including photocopying or any information storage and retrieval system without written approval by the publisher, except by reviewers who may quote brief excerpts in connection with a review or for private use.

Printed in the United States of America

CONTENTS

Contents v
Preface vii
Abbreviations xi

1. Frank Jack Fletcher

Task Force Commander 1
Early life 5
Continuous Combat 12
Legacy 17
Q & A 23
Wikipedia 34
Stormy Seas 43
Frank Friday Fletcher 52
Fletcher class DD 53

2. Till They Met Fletcher

Rendezvous 55
Pearl Harbor 58
Pacific Fleet 67
Aircraft 74
Wake Island 81
Java Sea, Map 83
Carrier Movements 85
Coral Sea 91
Midway 100
IJN Carriers 123
Guadalcanal 129
Savo Island 134
Hudson Myth 139
USN Aircraft 143
IJN Navy Planes 149
Eastern Solomons 152
Coast Watchers 155
Sub Attacks 156
Cape Esperance 158
Santa Cruz 159
Antiaircraft Light Cruisers 161
Naval Battle of Guadalcanal 162
Disarmament Treaty 165
Dud Torpedoes 166
Tassafaronga 171
Japanese submarines 172
Where are we? 174

3. Reference

Grand Strategy	177
Why Early Years	185
First Days of War	188
Atrocity	197
Ship Types	194
US Battleships	199
US Carriers	203
US Cruisers	206
IJN Battleships	209
IJN Cruisers	210
IJN Carriers	213
Pacific Commanders	216
Rank, Rate etc	217
Home Front, 1941	222
Relocation	228
Rest of the War	238
Operation Olympic	246
Speeches	248
Sources	254
About the Author	262

PREFACE

World War II was the defining event of a century. I had to study the war to understand what was the constant subject of the adults during my school years. This book represents decades of reading and the topics covered here include many tables of information that I wish I had access to while doing those studies. The book stresses the early war because the late war is well covered. At the end we had twelve million people in uniform and many wrote of their experiences. At the start of the war in Europe, the U.S. had less than one-half million in men uniform. Unfortunately, many of those on the front-line in the Pacific War were killed, so there were even fewer to survive to write of the early period.

The preponderance of late war writing tells of our grand victories in which we had superior weapons in vast numbers in fighting a weakened enemy. Early in the war we were equipped with the remnants of a depression era military of which politicians bragged of cutting the military budget to the bone and then a bit more. A lot of people died because of that thinking -- war is a come as you are party. We had two years of warning because of the war in Europe but we were still under-prepared. During that grace period, the U.S. authorized a lot of preparations by ordering new equipment and training programs that would have returned us to a reasonable level of defense. But actual delivery of these things and those people authorized had not yet arrived. For example, new draftees drilled with WWI or dummy guns to attack dummy tanks which were simply trucks with the word "tank" written on the side. No capital ships authorized after Pearl Harbor were able to be completed in time to participate in that war.

We were losing the War in the first year. Hitler had overrun Europe, except for Britain, and was attempting Russia and Africa. The Japanese fleet was running rampant in the Pacific – having destroyed the American Battle Fleet at Pearl Harbor, then captured SE Asia, the Philippines and the resource-rich East Indies necessary for their continued conquest of China. On their way they defeated the British Far East Fleet, the U.S. Asiatic Fleet, the Netherlands East Indies Fleet and sent the British Indian Ocean Fleet from Ceylon all the way to Africa minus a carrier and two heavy cruisers.

The Japanese fleet was armed with the top fighter airplane, they had superior torpedoes, night fighting capacities, and were with officers and men forged in several years of fighting in China. They had achieved their goals in the Pacific within five months and were in the process of extending those goals when the met Admiral Frank Jack Fletcher.

The Imperial Japanese fleet had never lost a ship larger than a destroyer or submarine. Fletcher stood in their way in an attack on Australia were they were bloodied for the first time and retreated. Though damaged, Fletcher raced to stem the next attack at the Battle of Midway and sank four of the Japanese big carriers. Let us point out that the Japanese started the war with ten battleships and ten aircraft carriers. After Pearl Harbor the U.S. had no battleships and three carriers in the Pacific. Always outnumbered and with the remnants of a depression era fleet, Fletcher fought the Japanese in three of the five carrier battles of the war. He sank six of Japan's carriers, the last one while defending the Marine beachhead at Guadalcanal. He never had more than three carriers to do this. He is an unsung hero of WWII and deserves to have his story told. This is why you are reading this book. Fletcher said that he always felt he was groping in the dark -- there had never been a carrier battle before he initiated the Battle of the Coral Sea in which ships of neither side sighted the other. He was writing the book on carrier operations and his native instincts were superior enough to make him the most successful admiral of the war. I hope that the groping in this book is adequate to tell his story.

Several people have written about Fletcher. There should be dozens to properly represent his importance. Regan, from up the road a piece from Fletcher's home in Iowa, wrote his biography. He also tries to pass off some of the things said by Morison. Morison is the author of the 15 volume official naval history of WWII that gives Fletcher the proverbial shaft. It seems that Fletcher refused to come out of retirement to help Morison write his books as other admirals did and they received favorable mention. Lundstom is a most exact and business-like writer of this period and recognizes that Fletcher was a key man who has not received the credit he deserves. When writing of the early war period, he documents in detail, as is his way, that everything Fletcher did had sound reason. I find myself torn between simply describing Fletcher's achievements and reacting with anger at the betrayal of historic honesty by Morison.

An avowed disciple of Fletcher is an Argentine consultant who writes on business decision-making – he sees Fletcher as perfection in decision-making under uncertainty. He has expanded my understanding and research by his knowledgeable questions. My personal thoughts are that Fletcher was a professional warrior, the man on the spot who was both competent and terribly lucky that he guessed right so often. However, for a leader to be consistently lucky requires something beyond chance. Napoleon said that he wanted lucky Generals. That also can be applied to Admirals.

This commemorative edition is in honor of the 125th anniversary of the birth of Frank Jack Fletcher in Marshalltown, Iowa, and is prepared for the Marshall County History Museum.

Jim Bauer
The Manorborn,
Marshalltown, Iowa
December 7, 2010

Vice Admiral Frank Jack Fletcher
arriving to accept the surrender
of the Japanese Northern Fleet

P.S. A note on the format of the book.
Much of this book is taken from the website,
    http://www.ww2pacific.com .
The webbish things have been removed but some may remain in spite of careful editing. On the website, the name of a ship that was sunk or damaged or newly commissioned are each in different colors, but these show up on paper as simply gray and have been changed to bold. The practice of surrounding semicolons, colons and dashes with spaces may carry through, in violation of what you learned in school, because of the limitation of monitors to clearly show these punctuation marks. The text font is intended to be Times New Roman, but the website uses Arial (without serifs) and not all changes may have been successful. Some tables may have had a column or two removed to fit on a page. There may be redundancy because most web pages are stand-alone topics. The website may have more information than can fit into these pages ; in addition, the website has more topics than could be included here. Any error you report will be corrected on the website immediately, which is not possible in book form. All errors are mine.

JLB

MAPS

Pearl Harbor Attack Fleet	58
Pearl Harbor Anchorage	58
Java Sea	83
Pacific 1942	84
Coral Sea	94

ABBREVIATIONS

These are used throughout the book without continual definitions. Some of the most used abbreviations are described here.

Ship Type. U.S. ships have a letter type description followed by a sequential number within the type. Example: USS Yorktown (CV-5) is the fifth ship in the series of fleet aircraft carriers.

- BB battleship, 14" guns or larger
- CA heavy cruiser, 8" guns
- CL light cruiser, 6" guns
- CLaa antiaircraft light cruiser, many 5" guns
- CM ex-US: merchant cruiser, heavily armed ship ; USN: minelayer
- CV fleet aircraft carrier, large and fast
  "V" is Navy speak for aircraft.
- CVL light fleet aircraft carrier, fast, built from cruiser hull
- CVE escort carrier, merchant hull w/flight deck used for convoy escort, aircraft ferry, anti-submarine.
- CVS Japanese seaplane carrier. Postwar, USN anti-sub carrier.
- DD destroyer, fast ship with 5" guns
- DE destroyer escort, smaller than DD, fewer 5" guns or smaller
- DM destroyer minelayer, converted destroyer
- PG patrol ship, gunboat
- PR river patrol gunboat
- SC sub chaser, for harbor defense
- SS submarine, USN.
- S older U.S. submarines, survivors were renumbered to SS series.
- U German undersea ship, U-boat
- I Japanese fleet submarine (very long range)
- RO Japanese coast defense submarine (shorter range)

Auxiliaries are supporting ships of the Navy not intended for fleet combat but are armed for self-defense.

- AVG aircraft carrier ferry; interim designation of CVE. Those built for Britain were BAVG.
- APV First designation for USS *Long Island,* a merchant hull with a flight deck, intended to relieve fleet carriers from ferry duty. This type was later designated AVG, then as a combatant, CVE.
- AD destroyer tender for resupply and repair
- AG general auxiliary, such as an HQ ship
- AH hospital ship
- AK cargo ship
- AKA attack cargo ship, carries landing craft.
- AM minesweeper
- AO oiler, an armed tanker

AP	troopship, often a converted ocean liner
APA	attack troopship, with landing craft
APD	fast troop carrier converted from a destroyer
AR	repair ship
AS	submarine tender – resupply and repair
AT	ocean tug
AV	seaplane tender – refuel, rearm, repair
AVD	seaplane tender, converted destroyer – refuel and rearm
IX	miscellaneous types

Ranks

FDR	President Roosevelt
CNO	Chief of Naval Operations
Adm	Admiral – any admiral or a 4-star
VAdm	Vice Admiral (3-star)
RAdm	Rear Admiral (2-star)
Capt	Captain
Cmdr	Commander
LCdr	Lieutenant Commander

Services

AAC	Army Air Corp – combat arm
AAF	Army Air Force
HMS	His Majesty's Ship (British)
IJN	Imperial Japanese Navy
RN	British Royal Navy
RAN	Royal Australian Navy
RNN	Royal Netherlands Navy
USAT	U.S. Army Transport
USMC	United States Marine Corps
USS	United States Ship (Navy)

Other.

AA	antiaircraft
a/c	aircraft
CAP	Combat Air Patrol ; fighter cover over fleet
TF	Task Force, a grouping of warships, essentially a fleet in the early war.
TG	Task Group, subdivision of a Task Force.
VF	Squadron of fighter airplanes
VS, VB	Scout or Bomber squadron. One of each on carrier; they used the same type of airplanes.
VT	Torpedo bomber squadron.

Division (2+ ships of the same type)
Squadron (2+ divisions plus tender)
Flotilla (2+ squadrons of destroyers or other small ships plus auxiliaries.

ADMIRAL FRANK JACK FLETCHER
Task Force Commander

Frank Jack Fletcher was born in **Marshalltown, Iowa**, on 29 April 1885. Appointed to the U. S. Naval Academy from his native state in 1902 ; he graduated from Annapolis on 12 February 1906 and was commissioned an Ensign on 13 February 1908 following the required two years at sea.

The early years of his career included service on the battleships *Rhode Island, Ohio,* and *Maine.* He served on survey ships USS *Eagle* and USS *Franklin.* In November 1909 he was assigned to USS *Chauncey* (DD-3), a unit in the Asiatic Torpedo Flotilla.

He assumed command of USS *Dale* (DD-4) in April 1910 and in March 1912 returned to *Chauncey* as Commanding Officer.

Transferring to USS *Florida* (BB-30) in December 1912, he was aboard that battleship during the occupation of Vera Cruz, Mexico, in April 1914.

For distinguished conduct in battle engagements ashore at Vera Cruz, he was awarded the **Medal of Honor.**

He became Aide and Flag Lieutenant on the staff of the Commander in Chief, U. S. Atlantic Fleet in July 1914. After a year at this post, he returned to the Naval Academy for duty in the Executive Department. His marriage to Martha Richards in February 1917 was for the rest of their lives; they had no children.

Upon the outbreak of World War I, he served as Gunnery Officer of USS *Kearsarge* (BB-5) until September 1917, when he assumed command of USS *Margaret* (SP-527). He was assigned to USS *Allen* (DD-66) in February 1918 before taking command of USS *Benham* (DD-49) in May.

He was awarded the Navy Cross :
> "For distinguished service as Commanding Officer USS *Benham*, engaged in the important, exacting, and hazardous duty of patrolling European waters and protecting vitally important convoys"

From October 1918 to February 1919, he assisted in fitting out USS *Crane* (DD-109) at San Francisco. Upon the commissioning of USS *Gridley* (DD-92) Commander Frank Fletcher was in command.. Returning to Washington, he was head of the Detail Section, Enlisted Personnel Division in the Bureau of Navigation from April 1919 until September 1922.

He returned to Asiatic Station, having consecutive commands of the USS *Whipple*, USS *Sacramento*, USS *Rainbow*, and Submarine Base, Cavite, P.I. He served at the Washington Navy Yard from March 1925 to August 1927; became Executive Officer of USS *Colorado* (BB-45); and completed the Senior Course at the Naval War College, Newport, RI, in June 1930. He went on to be graduated from the Army War College, Washington Barracks, in 1931. His experience and training qualified him for higher recognition.

He became Chief of Staff to the Commander in Chief, U. S. Atlantic Fleet in August 1931. In the summer of 1933, he was transferred to the Office of the Chief of Naval Operations. Following this assignment he had duty from November 1933 to May 1936 as Aide to the Secretary of the Navy, the Honorable Claude A. Swanson. He assumed command of USS *New Mexico* (BB-40), flagship of Battleship Division Three in June 1936. In December 1937 he became a member of the Naval Examining Board and became Assistant Chief of the Bureau of Navigation in June 1938.

Returning to the Pacific, between September 1939 and December 1941 he became Commander Cruiser Division Three; Commander Cruiser Division Six; and then concurrently, Commander Cruisers Scouting Force and Commander Cruiser Division Four. He was operating south of Oahu in USS *Minneapolis* when the Japanese attacked Pearl Harbor. Wake Island was threatened and he was in command of TF 14 in the attempted relief. He was in command of one of the two aircraft carrier Task Forces participating in operations in the Marshall and Gilbert Islands in February 1942 and was second in command during the Salamaua-Lae operations in March.

On 19 April 1942, he was designated Commander Cruisers, Pacific Fleet. This was the senior title available; Adm Pye had Battleships and VAdm Halsey had Carriers. He was in this command in May 1942 during the **Battle of the Coral Sea** which stopped the Japanese expansion -- *and saved Australia.*

In June at the **Battle of Midway**, he was Senior Task Force Commander, his flag flying in USS *Yorktown* (CV-5). It was in this battle that the Japanese suffered the first decisive defeat in three hundred and fifty years, restoring the balance of naval power in the Pacific -- *and saved Hawaii.*

Fletcher was promoted to Vice Admiral and awarded the Distinguished Service Medal

> "for exceptionally meritorious service as Task Force Commander, United States Pacific Fleet... during the Battles of the Coral Sea and Midway."

He commanded Task Forces engaged during the **Tulagi-Guadalcanal** landings on 7-8 August 1942 and two weeks later, the Task Forces in the ensuing **Battle of the Eastern Solomons** that repelled the enemy – *and saved Guadalcanal*.

In November 1942, he became Commandant, Thirteenth Naval District, Seattle, Washington, and Commander Northwestern Sea Frontier. He later became Commander, Alaskan Sea Frontier, with additional duty as Commander North Pacific Force and North Pacific Ocean Area. This was one of the three ocean areas under Nimitz. A task force under his overall command on 4 February 1944 made the first sea bombardment of the Kurile Islands. Determining the Kuriles were in a defensive posture, he felt free to interdict shipping in the whole northern Pacific. He returned to bombard the Kuriles in January 1945 claiming 30 ships destroyed. The same task force made the first penetration through the Kurile Islands into the Sea of Okhotsk on 3-4 March 1945. He was awarded the Distinguished Service Medal by the War Department for "

> ... his professional ability and able leadership in the vast wartime expansion and organization of naval installations in the North Pacific Area... between October 1943 and August 1945."

In September 1945, following the cessation of hostilities in the Far East, he proceeded to Mutsu Bay, off Ominato, Honshu, Japan, with sixty ships of his North Pacific Force for the emergency naval occupation of northern Japan where he accepted the surrender of the Imperial Northern Fleet. Included in his speech to his men:

> "Recalling the rape of Nanking, the treachery of Pearl Harbor, the Death March of Bataan, the murder, torture, and starvation of our comrades in arms, ours will not be an occupation in the Japanese manner. We have shown the Japanese and the world the superiority of our arms. We must now demonstrate to the world and the Japanese people the superiority of these standards of justice and decency for which we fought."

With the war won, he reported for duty as a member of the governing General Board, Navy Department, 17 December 1945. On 1 May 1946, he became Chairman, General Board, and continued to serve in that capacity until retirement in 1947.

Admiral Frank Jack Fletcher died 25 April 1973 and is buried at Arlington National Cemetery.

The Early Life of Frank Jack Fletcher

Frank Jack Fletcher was born April 29, 1885, and raised during his formative years in the home that his grandfather built at 202 Church Street, Marshalltown, Iowa. This was the second house built by George Glick after his family outgrew the first and when he became prosperous. Both houses stand today as museums to the family.

The house sits one block from the public square on which was being built the Marshall County Courthouse, dedicated 1886, and is still in use, and in the midst of the principle business district on Main Street. The 6,000 square foot, two-story brick house sits on a corner lot of 120 feet by 180 feet with a house of such substance as to become a hotel, the "Dodge House", soon after Jack went on to Annapolis and the house was sold. The stables across the backyard were almost the size of the house. Church Street is aptly named and Jack was raised across from the Stone Church, with a Methodist Church on the next block of the many churches on that street.

George Glick (1827-1910), his grandfather, immigrated to America at the age of thirteen with his parents, Johann Georg Glick and Katharina Geier from Otterberg, Bavaria, in May 1840 to Mansfield, Ohio. On graduation from Cincinnati Medical College in 1849, George and a professor, Eli Teegarden, went to the California gold fields as doctors and miners for five years. When he returned, he married Jane Ziegenfelder (1835-1911) of Pickaway, Ohio, Nov 1, 1855. Persuaded by relatives to take a half interest in a dry goods and grocery store, he moved to Marshalltown, Iowa, 1856.

Marshalltown had been founded in 1853 by Henry Anson. An 1860 census shows 30 buildings in Marshalltown. Population in 1875 - 4,384 ; 1885 – 8,298. By 1893 the city had grown to need a four-story high school.

In 1858, George opened a drug store and was appointed Postmaster. He ran the drug store with post office attached until 1870, speculated in land, helped build the town cemetery and was a leader in getting the railroad to run through the town in 1863. In fact, two rail lines intersect and remain as junction and repair yard to this day. As a result, George was elected a city councilman in 1868.

George and Jane Glick raised five children in a modest house at 201 East State Street they had built in 1859. That home is now the Glick-Sower Museum and was placed on the National Register of Historic Places in April 1993. They raised 5 children :
  Charles (1856-1914),
  Alice (1858-1945),
  Albert G. (1860-1917),
  Frank E. (1865-1881), and
  Fredrick E. (1872-1938).

Charles sold furniture and ran an undertaking business in 1893, and was Secretary of Lennox Industries. Alice married Thomas Jack Fletcher. Albert G. married Nellie Abbott, prominent in Marshalltown. Frank E. died in his teens. Fred E. in a lawsuit was said to be "with no means and no property".

In 1863 George Glick formed with Rev De Los and others the Marshall Cemetery Association for the Riverview Cemetery -- a 65 acre showplace to this day.

In 1864 he was treasurer of the school board and also in 1864 he helped establish the First National Bank of Marshalltown acting as cashier, vice president, and president for 18 years.

Dr. George Glick built a new, larger house for his family in 1870 where Jack Fletcher was raised until he entered the Naval Academy. The house is of such substance that when the family sold it about 1905 it became a hotel under the name of the Dodge House Hotel and the Wilson Hotel. The Historical Society of Marshall County purchased the building in the mid-1980s and opened it as the new museum and main office since 1987.

George Glick and two partners built Hawkeye Oil Mill, 1872, to process flax oil in which he both sold the seed and bought the crops; it was bought by Standard Oil which closed Marshalltown operations in 1878. The building is now Diamond Vogle Paint factory.

With four partners, he organized the Marshalltown Public Library in 1879. This became one of 2,509 Carnegie Library buildings in 1902.

He was a co-founder in 1879 of The Bankers Life Association (now Principle Financial Group), to provide low-cost life insurance for fellow bankers and their employees and served as its first medical officer.

With four partners he formed Marshalltown Sugar Refining Company 1880, known as the "Glucose Company", to process corn, cane, and potatoes – it was purchased by Firmenenich, Germany, employing 400 workers – it closed in 1891 for pollution problems.

He persuaded the Iowa State Assembly to build the Soldier's Home, now the Iowa Veterans Home, which opened in 1887 on 160 acres and is still a major factor in Marshalltown and Iowa.

He moved to Des Moines in 1893 but retained a vice presidency of his bank. Glick Elementary School is named in his honor, built in 1902 to serve the third ward and had eight teachers. George Glick died in 1910 in Marshalltown and is buried on Founder's Hill in the cemetery he started.

Thomas J. Fletcher, (1842-1929), his father, was born in Indiana and moved to Oskaloosa, Iowa, as a child. He served in the Civil War, joining 25th Iowa 9th Infantry at Pella. Mustered out July 1, 1865, at Louisville, KY. He moved to Marshalltown and was made Cashier of the First National Bank in 1877. In October of that year he married George Glick's only daughter, Alice, age 18. They first lived at 9 South Third Street and then a large house at 106 South Second Avenue, where is now the children's wing of the Episcopal Church. In 1884 they moved into her family's home and raised three children there : Emma Belle (called Belle) (1881-1966) trained as a teacher and moved to Hollywood, California ; George G. (1882-1947) moved to Des Moines; and Frank Jack (called Jack) (1885-1973).

T.J. was Cashier, First National Bank 1877-1882, succeeded by A. G. Glick.

In 1882 he was treasurer of the Glucose Company.

Thomas was the co-founder with Frank Letts of the Letts-Fletcher Wholesale Grocery in 1884 where Tom was treasurer. The building is located at the other end of the block from the Glick-Fletcher home. Tom ran operations and was treasurer until 1889 when they expanded to Carroll, Iowa, and he sold his interests. The Wholesale Grocery operated through 1946. The building became Iowa Plumbing Wholesale in 1963 and a grant was received in 2010 for it to be converted into retirement apartments.

He rejoined the Bank as Cashier in 1891 for a further 16 years until the First National Bank was sold by the Woodbury brothers to City National in 1907.

Tom Fletcher and Frank Letts founded the Marshalltown Public Library established in 1898 and built with Carnegie Library funds in 1902.

About 1893 Thomas was Local Treasurer of Iowa Central Railway, a 49 mile, 67 locomotive, freight line.

Soon after retiring from the bank, Tom and Alice went to live with their daughter, Belle, in Hollywood, CA, and later returned to Marshalltown. T.J. died June 30, 1929. Alice Glick Fletcher lived to see her son's victories in WWII. She died Oct 21, 1945. Both and their daughter, Bell, are buried on Founder's Hill in Riverside Cemetery in Marshalltown.

Frank Friday Fletcher (November 23, 1855 - November 28, 1928), uncle, was born in Okaloosa, Iowa, and went to Annapolis, class of 1875. He became a specialist in new weapons and was a LCdr at the Torpedo Station, Newport, RI, when Jack went off to Annapolis. Rear Admiral Frank Friday Fletcher earned the Medal of Honor for his command of the occupation of Veracruz in 1914 and became Admiral of the Atlantic Fleet before World War One. He served on the General Board until 1919 and was twice recalled for temporary active duty.

Alice Glick Fletcher (Oct 22, 1858 - Oct 21, 1945), mother, was almost as an enthusiastic booster of Marshalltown as her father and husband. She taught at the Industrial School for underprivileged children. She chaired the library committee. Was active in woman's rights. Helped form the Iowa Federation of Woman's Clubs, was State Secretary in 1901 and President in 1903. She promoted the traveling library (now inter-library loan system). She was a charter member of the Hawthorn Club, Twentieth Century Club, and Marshalltown Country Club. She spent some time living with her daughter, Belle, in Hollywood, California. Alice died in 1945 and is buried in Marshalltown's Riverside Cemetery with her family.

Frank Jack Fletcher. (Apr 29, 1885 - Apr 25, 1973). Jack was raised in a household devoted to public service and entrepreneurship and familiar with politics. Young Jack attended the public school, played football and graduated from Marshalltown High School among a class of 29, Jack was appointed to Annapolis in 1902 by Senator Jonathan Dolliver. The Fletcher family home was sold soon after Jack left and the family moved to 5 S. Fifth Avenue, now part of the hospital parking lot.

Jack is only known to have returned once to Marshalltown, for the funeral of his mother in October, 1945.

His Neighborhood

An open town square was the center of Marshalltown when Frank was born a block away on the corner of 2nd Street and Church Street. The present county courthouse was built on that square before he reached school age. The block on in which he lived had only seven houses, his and another facing Church Street, the others facing Main Street. Across the street was the "stone church" which had a varied history. By the time he was ten years old the center of town had grown, the church was Universalist, the Odeon Opera House was built next door, across Second Street in 1888; a steam plant ran heat to the Courthouse and the Odeon ; a carriage works was built between his house and the Courthouse, and Doctor Gets had established a Surgical Hospital on Third Street. By the turn of the century, when Frank turned fifteen, the Stodart Hotel and the Letts-Fletcher Wholesale Grocery, then Wholesale Supply, now Senior Residences, were built on the property to the back of the Fletcher house, facing Main Street, and three shops where built diagonally across his block on Main Street -- a tailor, a bakery, and a confectionery.

The Glick-Fletcher House (now Marshalltown Historical Museum) had two families of interest: his grandparents, George and Jane Glick, and his parents, Thomas and Alice Fletcher, with his sister Emma Belle (1881-1966) teacher, later to Hollywood; his brother George G. (1884-1947) later retired to Des Moines; and Frank Jack (1885-1974) who became a naval flag officer.

The "stone church" across the street had a varied history. The Episcopalians had met at Woodbury Hall from 1863 until1870 when they bought the former Lutheran Church, the first church built in Marshalltown, 18 E. State Street (then River Street), and named it St Mathew's Chapel. That building was sold to the Swedish Lutheran Church in 1871 and construction started of a stone building on the current church site. St. Matthew's Church was completed in 1872. During the depression of 1878 they were forced to sell off parts of furnishings and finally turn over the deed to the mortgage holder in 1880. The building was sold to Unity Universalist which worshiped there for 15 years before leasing to the People's Church in 1897 and to the Church of Christ Scientists in 1899 until their beautiful prairie style building was completed in 1903.

The Episcopal congregation changed their name to St. Paul's. A new frame building was built at 107 North Center Street by the Ladies Guild and called the Guild Hall. It expanded and evolved into St. Paul's Church. In 1903 the original stone church was repurchased from the Universalists, renovated and enlarged as the present Saint Paul's Episcopal Church.

The Methodist Church down the street at First Avenue and Center Street was established in 1863. A new location was built at Second Street and West Main and is still the First United Methodist Church. The old building became a theater.

The Odeon Opera House, at 13-17 S. Second Avenue, across the street beside Fletcher's home, was built in 1888 and was where some of the biggest names in live theater came. It began showing movies in the 1920s, becoming all movies by 1935. It burned in 1942, was rebuilt and reopened in 1943 as a beautiful 756 seat movie hall with a large performance stage where Wallace Beery hosted the opening and where Tim Holt, Walter Brennan, and Ruth Warrick performed. With the advent of television, business waned and the building was torn down in 1957 and remains a parking lot today.

Fisher Governor. In 1880 William Fisher was town waterworks manager. In a major fire, he had to regulate water pressure by hand for many hours. He invented a constant pressure pump governor and started to build them for other waterworks around the country. Early employees operated belt-powered machines to the light of tallow candles for $13-a-week. The first product catalog was hand-written. He incorporated as the Fisher Governor Company in 1899 with a total of $30,000. Fisher Controls is now a billion-dollar business.

Other than churches. In 1900 the seventeen saloons in Marshalltown were temporarily closed by the Supreme Court until new licenses were approved.

Annapolis

Frank went off to the Naval Academy at Annapolis, Md, in 1902. Each Senator and Representative may recommend two candidates for midshipmen each year. The candidate must be between sixteen and twenty years of age at the time of the examination for admission. Candidates must be physically sound and over five foot two and at least 105 pounds at age 16 and, these numbers increase with age. The entrance examination is on Grammar, Geography, U.S. and World History, and Arithmetic. If the candidate passes, he immediately goes to the Academy. The course of study is six years – four years at the Academy and two years at sea followed by an examination for final graduation and assignment to one of the lower ranks of officers of the Navy or Marine Corps.

Frank Jack graduated 26th in his class of 1906 with high marks in conduct, efficiency, ordinance, and seamanship.

During the four years at Annapolis Naval Academy, a midshipman develops a passing or close acquaintanceship with members of seven classes, in his case, the classes of 1903 through 1909. This group was to comprise much of the naval leadership during World War II. Twenty two of 116 graduates of his class achieved flag rank. Names from that seven year period that earned fame in the War include :

- Seniors, class of '03 when he arrived were : Harold Stark, CNO ; Walter Anderson, BB Div4 ; Samuel Robinson, Procurement
- Juniors '04 : Husband Kimmel, CINCPAC ; William Halsey, Third Fleet
- Sophomores '05 : Chester Nimitz, CINCPAC ; Royal Ingersoll CINCLANT; Fairfax Leary, Eastern Sea Frontier ; Albert Church, flag officer ; John Newton, DCINCPAC ; William Furlong, Mine Force
- Fletcher's class of 1906 : Robert Ghormley, SoPacArea ; William Calhoun, ComSrvPac 1939-45 ; Leigh Noyes, Ch. Inspections ; William Glassford, ABDA ; Frank Jack Fletcher, Task Force Commander ; John Towers, DCINCPAC ; Milo Draemel, flag officer ; Isaac Kidd, Battleship One, died USS *Arizona* ; John McCain, BuAir ; Aubrey Fitch, NavAirSoPac.
- The class behind him, '07 : Robert Theobald, TF 8 ; Raymond Spruance, Fifth Fleet ; Kent Hewitt, Europe ; Richard Edwards, DCNO ;

Daniel Callaghan, TF Commander ; Norman Scott, TF Commander ; William Purnell, ChStaff Asiatic ; Patrick Bellinger, Atlantic Air
- Two years behind, '08 : Richmond Turner, Amphibious ; Thomas Kinkaid, Seventh Fleet ; Arthur.S. Carpender, SWPac ; John Hoover, Central Pac ; Willis Lee, Battleships ; Alan.G. Kirk, Mediterranean ; Lloyd Stark, Gov Mo.; Marc Mitscher (grad '10), Fast Carrier Force
- Entered when he was in his senior year '09 : Theodore Wilkinson, ONI ; Franklin Valkenburgh, Captain, *Arizona ;* Jesse Oldendorf, Battleship Squadron ; Jonas Ingram, CINCLANT

During World War I, Fletcher captained the destroyer *USS Benham* operating on convoy escort and patrol duty in British and French waters for which he was awarded the Navy Cross. He married Martha Richards in February 1917. She was the daughter of Gertrude M. and Walter B. Richards of Kansas City, MO. Martha Fletcher passed away September 14, 1974, in La Plaza, Maryland, at the age of 79, a year after her husband of fifty-six years.

Early in the war, Marshalltown had a war bond drive in his name, Sept 5, 1942, complete with a band concert on the Courthouse lawn. Such concerts are a usual summer event on the Marshalltown courthouse lawn to this day.

Continuous Combat : Dec 7, 1941 to Sept 30, 1942.
Commander of Heavy Cruiser Force 6 at Pearl Harbor.

Half of Cruiser Six was in drydock getting upgrades in anticipation of the coming fray. The other two were at sea.

Minneapolis (CA-36) departed Dec 5, 1941, to operate south of Oahu. December 7th, searches for Japanese fleet. This force is sighted and reported as an enemy aircraft carrier ; bombers are sent which recognize the U.S. ship. Returned to Pearl Harbor on Dec 10.

Astoria (CA-34) sortied December 5th with TF 12 built around *Lexington* (CV-2, RAdm Newton) to deliver aircraft to Midway Island. It was too late. The ferry mission was canceled with orders to search southwest of Oahu "to intercept and destroy any enemy ship in the vicinity of Pearl Harbor. . . ." They reentered Pearl Harbor on 13 December.

We don't know which of his ships Fletcher was with, but he took the heavy cruiser division to sea on the 16th to rendezvous with and screen an emergency convoy to Wake Island.

Attempted Relief of Wake Island

The Japanese invasion of Wake Island on December 11th is repulsed by Marines on the 7 sq mile atoll. Shore battery gunfire sinks destroyer *Hayate* and damages four others. USMC F4F Wildcats bomb and sink destroyer *Kisaragi* and strafe and damage two light cruisers. Following the abortive assault, the Japanese bomb daily with flying boats in pre-dawn raids and later in the day, land attack planes bomb Wake.

The American public is encouraged by this victory.

On the 14th, VAdm Brown, TF 12, with *Lexington* and escorts sails for the Marshall Islands, to disrupt the suspected source of the attacks on Wake Island.

Next day, a relief convoy around large seaplane tender *Tangier* departs Pearl as carrier *Saratoga* arrives from the West Coast to refuel.

The day after, Fletcher's cruiser division escorts *Saratoga* to catch up with the convoy that is also to return 1,145 civilian contractors. The Japanese Pearl Harbor Force detaches carriers *Hiryu* and *Soryu*, with heavy escorts to reinforce a planned second attack on Wake Island. Japanese naval land attack planes continue to bomb Wake.

Admiral Kimmel is relieved on the 17th by SecNav Knox by VAdm. William Pye, Commander of the no longer existent Battle Force, pending the arrival of Admiral Chester Nimitz.

Next day Pye orders *Lexington* north to join with *Saratoga*.
On the 20th, Halsey's TF 8 with *Enterprise* proceeds to waters west of Johnston Island to cover operations of the two other carriers and Hawaii.

On the 21st and 22nd, planes from carriers *Soryu* and *Hiryu* and land planes bomb Wake. The last two flyable USMC F4Fs intercept the raid. One is shot down, the other is badly damaged. Fletcher's *Saratoga* Task Force is ordered to wait for *Lexington* TF. Fletcher refuels his destroyers for the attack while waiting.

On Dec 23rd two small Japanese transports are intentionally run ashore to facilitate landing of naval troops. Planes from the two carriers as well as a seaplane carrier *Chitose* provide close air support for the invasion. Wake Island is captured after a gallant resistance offered by the garrison.

Uncertainty caused by the presence of Japanese carriers and reports that Japanese troops have landed on the atoll compel VAdm Pye to recall the relief mission.

During the return *Saratoga* flies off 20 USMC F2A Buffaloes to Midway. *Tangier* disembarks defensive troops and a ground echelon.

January 1, 1942 . Fletcher takes command of Task Force 17 centered on USS *Yorktown* (CV-4) returning to the Pacific Fleet after having been transferred eight months earlier for on service on Neutrality Patrol in the Atlantic. His first task is to escort troopships carrying Marines to the South Pacific which allows him to learn aircraft carrier operations. He raids Japanese held islands on the way back to Pearl Harbor.

Battles in Java Sea.
East Indies are lost

On the far side of the Pacific, the Japanese rapidly moves down the East Indies from their earlier conquests in Indochina and the Philippines and take on
Jan 10 - Borneo, Jan 23 - Bismarck Archipelago (Rabaul), Jan 24 - Celebes Feb 9 - New Guinea, Feb 16 - Sumatra, 19 Feb - Bali and make an air raid on Darwin, Feb 20 - Timor.

A joint task force of American, British, Dutch, Australian cruisers and destroyers under Dutch RAdm Karel Doorman attempts to attack invading troopships, with some success at Macassar Strait, Jan 24, but the fleet was lost in battles around Java : Badoeng (Lombok) Straits, Feb 18-19; Java Sea, Feb 27; and Sundra Strait, Feb 28.

Burma and the Philippines are captured in April and the Japanese Navy raids Ceylon and along the eastern coast of India which pushes the British fleet out of the Pacific and Indian Oceans back to Africa.

Battle of the Coral Sea – May 4-8. 1942
Japanese Expansion Stopped

Fletcher's *Yorktown* Task Force 17 was dispatched from Pearl Harbor to Australia, Feb 15, arriving about Feb 24. Fletcher's force joined under VAdm Brown's *Lexington* TF 11 to make air attacks from near Port Moresby in Papua, over the mountains to Salamaua and Lae on the north coast of New Guinea, on March 10 where they sink four ships, damage 13 others.. A U.S. fleet base of operations was established in the New Hebrides March 18. Fletcher's task force patrolled the Coral and Solomon Seas alone to protect shipping to Australia.

Other Imperial troops started taking the Solomon Islands with Bougainville, March 30, and down the island chain to Tulagi, opposite Guadalcanal, May 3. On May 4th Fletcher attacked this new seaplane base in the Solomons. His TF 17 is then joined by Fitch's *Lexington* TF 11 to head off a Japanese troop landing at Port Moresby on the south side of Papua New Guinea. This invasion would give the Japanese control over the seaways to Australia, even to threaten Australia itself. Meanwhile, a second Japanese fleet, with two fleet carriers, arrived to search for American carriers. Fletcher sent his planes to attack the invasion fleet and destroyed the supporting carrier: "scratch one flattop"; and assigned his Australian cruisers to block the occupation fleet while he held off the Japanese carrier fleet. The next morning, the naval air fleets found and attacked each other simultaneously. One enemy carrier was damaged and virtually all of the 121 IJN carrier planes destroyed. However, both U.S. carriers were damaged with a loss of 76 planes. The invasion force retreated ; both enemy fleet carriers were out of service until after the Battle of Midway. At the conclusion of the fighting the U.S. had won, but a few hours later, gasoline fumes on *Lexington* exploded. The "Lady Lex" had to be abandoned. *Yorktown* returned to Pearl Harbor for repairs.

The final numbers in the battle were in favor of the Japanese who lost a small carrier while the U.S. lost a large one. However, the Japanese attack and occupation of Port Moresby was thwarted and the Coral Sea remained a buffer for Australia and allowed for the continued U.S. naval buildup in the New Hebrides. And, two enemy big carriers were damaged so as to not be available for the next battle.

Battle of Midway – June 4-6, 1942
Major U.S. Victory

The Japanese did not feel they could continue to encounter the U.S. fleet in isolated instances, there were too many risks and false alarms that distracted the fleets, required extra support to each troop movement, and required a home fleet to protect the main islands of Japan. The strength of the Japanese fleet was far superior to the U.S. Pacific Fleet and a plan was drawn up to bring out the entire U.S. fleet to be defeated once and for all. Admiral Yamamoto's plan included a diversionary taking of Aleutian islands and a frontal assault on Midway with over 200 ships. There were two carrier strike fleets (Dutch Harbor and Midway), three invasion fleets (Midway, Kiska, and Attu). Included were 11 battleships, four fleet carriers, and five smaller carriers in covering, supply, and screening forces. From coded messages intercepted, the U.S. fleet knew of the Japanese attack and waited in ambush. All four enemy strike carriers were destroyed ; the *Yorktown* was again damaged and later sunk by a submarine while under tow. But, the excessive Japanese power in the Pacific had been broken.

Invasion of Guadalcanal – Aug 7, 1942
A "Shoestring" effort to stop the advances.

Newly promoted Vice Admiral Fletcher was the senior officer and commander of three carriers to escort and defend the troops to the landing. U.S. Marines landed at Tulagi and Guadalcanal to protect the Australian shipping lanes. Tulagi was now an established seaplane base and strongly defended, but construction on Guadalcanal was undefended. The Japanese air response was immediate with bombers disrupting the unloading and damaging a transport and two warships. Fletcher's carrier aircraft provided close air support and defended from air attacks. When the beaches were secured, he went back to his primary mission to prepare for the expected counter-invasion with strong Japanese aircraft carrier support.

Battle of Savo Island Island – Aug 9, 1942
Major U.S. Disaster

The Japanese response to the invasion of Guadalcanal was to immediately send a cruiser task force from Rabaul to interrupt the landing. This force arrived the second night and destroyed four heavy cruisers, *Quincy, Vincennes, Astoria, Canberra,* of the screening force in 32 minutes and a previously wounded destroyer, *Jervis*. The Navy lost more men (1,270 killed, 709 wounded) on that one night than the Marine Corps did in the whole Guadalcanal campaign. Fletcher was in the process of withdrawing his carriers after the successful landings and could not give chase. With the cruisers sunk and the carriers gone, the U.S. transports were exposed. To save his ships, RAdm Turner retreated with several ships still unloaded and the Marines were left to defend their occupation.

Battle of the Eastern Solomon Sea – August 24, 1942
Air Fight.

The second naval battle in the Solomons , the third carrier battle of the war, took place on August 23-25, 1942, and was fought with aircraft -- eight carriers were involved. Henderson Air Field on Guadalcanal had become operational on the 21st with Marine Fighter Squadron 223 and Scout Bombing Squadron 232, delivered by the auxiliary aircraft carrier *Long Island* on Aug 20. Adm Yamamoto gathered 58 warships including 4 BB, 2 CV, 3 CVL, 13 CA, 3 CL, and 30 DD with the mission to smash American naval forces in the SW Pacific and convoy three troopships to reinforce Guadalcanal. VAdm Fletcher's TF 61 of 28 ships had one battleship, *North Carolina* ; 3 big carriers ; 5 CA , 2 CL ; and 18 DD.

Both sides fought cautiously. Fletcher, although outnumbered two to one, attacked first. A Japanese small carrier was sunk. The two U.S. carriers engaged were damaged : *Enterprise* and *Saratoga*. The *Wasp* task group had been detached on refueling rotation and did not participate. With aircraft depleted, the Japanese fleet withdrew with Yamamoto no doubt remembering Midway and wondering where was the other U.S. carrier? The next day, a Japanese light cruiser, a destroyer, and a troopship were sunk by Marine planes from the newly opened Henderson Field on Guadalcanal.

At the end of September Fletcher returned to the States after almost nine months of continuous combat and the fleet organization was shuffled from defensive to the offense.

In the **Battle of the Coral Sea**, he saved Australia, sank one light carrier, and damaged two of the six enemy carriers that had attacked Pearl Harbor so that they could not join in for the **Battle of Midway**, where he destroyed the other four of Japan's large fleet carriers. He stopped the major enemy thrust to retake Guadalcanal in the **Eastern Solomons** while removing another enemy carrier before returning home.

He was assigned as Commander Northwestern Sea Frontier, Nov 1942, to lend his prestige to the defense of the U.S. mainland from a northern attack.

The Northwest Sea Frontier

The Japanese had taken two Aleutian Islands while Fletcher was fighting at Midway. These islands are on the most direct route from Tokyo to Seattle. The public perceived an invasion from the North and the Northwestern Command was in disarray. Fletcher had the prestige and diplomacy to straighten things out between the various U.S. armed services, civilians, and Canada. When this was accomplished, Fletcher was given command of the entire North Pacific, but in this theater, where the major enemy was the weather, he had to content himself with transporting aid by the western route to Russia : 6,400 planes, 149 frigates and other small ships, and trained 8,700 Russian sailors. He was able to raid the Kurile Islands in 1944 and again in 1945. He accepted the surrender of the Imperial Japanese Northern Fleet in Mutsu Bay September 1945.

Legacy

Until recently, two books have defined Fletcher. Morison's *History of U.S. Naval Operations, Rising Sun* tends to dismiss Fletcher's accomplishments. Regan's *In Bitter Tempest: The Biography Of Admiral Frank Jack Fletcher* is good, but feels compelled to defend Fletcher from Morison's criticism. There is no need to apologize for battle losses in stopping the runaway advances of the enemy -- he lost two carriers but while sinking six -- pretty good results. A new book, Lundstrom's *Black Shoe Carrier Admiral,* goes meticulously into the facts and results of Fletcher's battles. And one waiting publication praises Fletcher, del Castillo's *The Days of Fletcher*.

Fletcher's sinking of six enemy carriers make him the most successful admiral of the Second World War -- Pacific and Atlantic. That he did this with the remnant of a small depression era fleet is what is so remarkable. A historian might note that he was the most successful admiral of World War I and World War II in sinking enemy capital ships.

He was appointed to the Navy's General Board in 1946 and retired as Chairman of that governing board in 1947 with the rank of full Admiral. Jack and Martha enjoyed life of retirement and entertaining on their historic county estate, Araby, in Maryland.

Frank Jack Fletcher died on 25 April 1973 at the Bethesda Naval Hospital in Bethesda, Maryland. He is buried in Arlington National Cemetery.

USS Fletcher (DD-992) is named in honor of Admiral Frank Jack Fletcher.

Fletcher Deserves More Recognition than He has Received.

Fletcher was competent and he won his battles. The U.S. Pacific Battle fleet, the British Far East Fleet, U.S. Asiatic fleet, the Dutch East Indies fleet and the British Indian Ocean fleet were all destroyed by the Imperial Japanese Navy without-loss-to-themselves until they encountered Fletcher.

Edwin Layton, in his book, *And I Was There*, gives an inside story at headquarters at Pearl Harbor. Layton was chief of intelligence for CINC Pacific. Layton saw Fletcher as old line Navy with no air experience who has a dim understanding of Layton's field of radio intercept analysis. He points out some facts that seem to indicate that Fletcher was handicapped by misimpressions at HQ and in Washington.

Adm Kimmel was Pacific Commander in Chief at Pearl Harbor. He immediately laid plans for attack against the Japanese including setting a trap at Wake Island for the aircraft carriers that had destroyed his battleships. The Marines on Wake Island had successfully fought off an invasion force on Dec 11, sinking two destroyers. Kimmel expected the Japanese to try again, this time with aircraft carrier support. He selected, over more senior admirals, RAdm Fletcher to head the Wake Island relief portion on Dec 15. The plan consisted of all three aircraft carrier task forces. *Lexington* (VAdm Brown, TF 11) was to make a raid in the Marshall Islands, south of Wake, from where the invasion troops would come ; *Saratoga*, RAdm Fletcher, TF 14, aboard cruiser *Astoria* (CA-34), to escort *Tangier* (AV-4) with relief supplies ; and *Enterprise* (VAdm Halsey, TF 8) to stand guard west of Johnston Island.

Washington decided Kimmel was to be replaced by Adm Nimitz, but Kimmel was relieved prematurely (and later made a scapegoat) with an interim commander, VAdm Pye, the Pearl Harbor battleship force commander, who botched the plan. Pye, having already lost his battleships, was mainly concerned with conserving the carriers to turn over to Nimitz. In addition, messages from Washington said Hawaii was more important to defend than Wake Island. Pye recalled Brown to join Fletcher and had Fletcher wait for Brown over 400 miles from Wake Island, out of air support range. When the Japanese detached two aircraft carriers from the Pearl Harbor raid to support the invasion of Wake Island on Dec 23, none of the three U.S. carriers were able to assist. Pye sent contradictory orders in rapid succession. In a day he told Fletcher to attack, then to send in *Tangier* alone, a half hour later told Fletcher to evacuate Wake, and in another half hour told him to withdraw. Later, Adm King, Chief of Naval Operations, in Washington, is reported to have blamed Fletcher for the loss of Wake Island.

Nimitz had been ChBurNav (later renamed Bureau of Personnel) in Washington. He took over the Pacific Fleet on Dec 31, retained Kimmel's staff, including Layton and Fletcher, and set to work to most effectively block Japanese expansion, establish a secure supply line to Australia, and to defend the Midway-

Pearl Harbor line. It was a time of confusion, of impossible demands from Washington, and of prioritizing. With insufficient strength to reach the Philippines, those reinforcements were diverted to Australia ; established a goal to extend the Hawaii-Samoa line to Fiji ; and planned raids to keep the Japanese away from Midway-Pearl Harbor.

Marcus, Wake, the Marshalls (Kwajalein) and Gilberts (Tarawa) form a line of advanced holdings just west of the International Date Line that allows the enemy an early warning of an approaching U.S. fleet. The later two lie near the shipping lanes to Australia. This eastern line complemented the major Japanese defensive line of the central Pacific bases in the Marianas (Saipan), Caroline Islands (Truk) and Bismarcks (Rabaul).

Fleet in Being. This is a technical term that means, just by its existence, a fleet imposes defensive measures and extra support for offensive operations on the opposing force. Example : As long as Germany had a fleet, the British had to maintain a battle force in readiness, to provide escort to every maritime activity, to plan attacks and defense, and in general tie up large forces just because they never knew when the enemy warships might sail.

The Japanese, having destroyed the U.S. Pacific Battle fleet, sunk the British Far East fleet, defeated the ABDA fleet, and expelled the British Indian Ocean fleet thereby required the U.S. to play a defensive role to retain a "fleet in being". The Japanese had to retain a large fleet in home waters, to send fleets to investigate every reported sighting, to provide strong escorts, and to always remain wary of an attack. The U.S. raids on Kwajalein, Makin, Wake, Lea, and Tulagi, although doing little direct material damage, plus Doolittle's raid on Tokyo, all involved moving the few available ships around the Pacific showing that the U.S. Navy still had teeth.

This forced the Japanese to insist on one big battle to eliminate the remaining U.S. fleet -- the Battle for Midway. The U.S. had to conserve their ships to maintain the threat of the American fleet; they could only engage in battle when the odds of damage to the enemy exceeded the risk to their own ships. Fletcher was the admiral at sea that had to execute this policy and to make the decisions to interpret this policy on the spot.

While *Yorktown* patrolled the Australian sea lanes, an Allied reconnaissance plane reported Fletcher off Rabaul. This was taken by CNO Adm King as an offensive by Fletcher. In fact, the sighting was of a Japanese force sailing to join their Port Moresby attack fleet. When Fletcher reported he was approaching Tongatabu for replenishment, Washington's excitement turned to disappointment. Fletcher had been following orders to patrol and attack when conditions were favorable. He rushed to intercept the Japanese probe but it was not found. Once again, Fletcher was innocently considered a disappointment by those in Washington who expected a quick victory over the slanty-eyed toymakers by the weakened, depression-era American fleet.

Critique.

Fletcher has been criticized for:
 Foul-up at Wake Island.
 Less aggressive than Halsey
 Losing the *Lexington* at Coral Sea, Adm King's former flagship.
 Losing a second carrier, *Yorktown*, at Midway, seemed unlucky.
 Surprise at Savo Island, losing an entire supporting fleet without retaliation.
 Leaving troops undefended at Guadalcanal ; Marine Corps incensed.
 Having *Wasp* out of the action at East Solomons.

Observations in Response.
* We have seen in Wake Island Relief that RAdm Fletcher was following orders; the problem was VAdm Pye's conflicted indecision caused by the premature removal of Kimmel. Pye had been told by Washington that Wake was secondary to the defense of Hawaii. However, after the fact, Roosevelt is said to have considered the fall of heroic Wake Island a worse blow than the bombing of Pearl Harbor.
* Fletcher had been appointed by Kimmel, and was on his shortlist of officers for that job, making Fletcher suspect by Adm King, the new Chief of Naval Operations in Washington
* VAdm Halsey was senior to Fletcher and made pithy remarks.
* RAdm Fletcher was not a public speaker and attracted less press coverage.
* Halsey raided Kwajalein, Marcus and Wake, escorted *Hornet* on the Doolittle raid of Tokyo -- all making for popular press.
* Fletcher's sinking of **Shoho** was the first destruction of a major Japanese ship in the war. And the enemy invasion force was turned back. The secondary explosion on **Lexington,** under Fitch, was a fact of battle, witness how few bombs sank four Japanese carriers at Midway. And *seven* English aircraft carriers in 1939-42.
* Fletcher's attacks sank all four Japanese fleet carriers at Midway: **Akagi, Kaga, Hiryu,** and **Soryu;** and was withdrawing for the night when he released RAdm Spruance for independent operations. When information was released to the press, Spruance was then commanding the remaining carriers and was reported erroneously with credit for the Navy victory. Later, when Spruance became a famous admiral, his biographers came back to his first public engagement and further credited him with victory in Fletcher's battle.
* *Yorktown* was the only carrier attacked at Midway and was finally sunk by a submarine two days later while under tow.
* The plan at Guadalcanal called for the carriers to be removed from the confines of the island when the landing was secured and to defend at sea from an expected counter-attack. Fletcher's primary responsibility was to fight enemy

carriers. The landing at Guadalcanal was <u>unopposed</u> and the Marine goal to capture the airfield was completed the first day. Marines had finished fighting at Tulagi on the 2nd day. With the landing secure, Fletcher went to prepare to fight carriers.

* Savo Island was a disaster. The USN was unprepared to fight a night action; the IJN were expert. However, the enemy withdrew without attacking the transports -- the defense was accomplished.
* Fletcher's carriers were out of range to punish the enemy cruisers which embarrassed desk-bound officers already shamed for being unprepared for night fighting.
* The Japanese could not land reinforcements until ten days later, and then with only 1,000 men -- who were wiped out by the 11,000 Marines.
* RAdm Turner was delayed in unloading and made the decision to save his transports from air attack and withdrew from Guadalcanal.
* All Marine accounts mention Fletcher's withdrawal ; none mention that it was that same Fletcher who protected them with those same ships when the Japanese Combined Fleet came calling two weeks later intent on conquest.
* Several historians repeat Morison's accusation of Fletcher's "leisurely" refueling. Others recognize this as preparation to engage the enemy whenever an opportunity for combat comes. Fletcher had enough fuel to stay near Guadalcanal, but not enough to fight a battle. Refueling at sea was a new technique, not perfected until mid-war.
* Fletcher refueled his carriers in rotation ; otherwise all might have been out of action at the same time and the U.S. would have lost the Battle of Eastern Solomons – and with it the entire Guadalcanal, Solomon Islands beachhead and again threaten Australia. Instead, Fletcher sank his sixth carrier, **Ryujo**, and the Japanese invasion fleet was repulsed from Guadalcanal.
* Fletcher was at sea carrying out impossible orders ; not in port "communicating".
* Fletcher's later orders were to escort and patrol the shipping route to Guadalcanal. This kept his carriers in a limited area south of that island and an invitation to submarine attack, an area later called Torpedo Junction, where his flagship *Saratoga* was damaged and he returned to Pearl Harbor. His successors lost **Wasp** with *North Carolina* damaged ; then **Hornet** was sunk All within two months, with no further enemy carriers damaged or destroyed.

Fletcher had also lost two carriers -- but he had sunk six enemy ones.
His successors lost two carriers, but sank none. Pretty good odds for Fletcher.
Of this heroes are made.

It has been suggested that if Halsey had been at Midway, rather than the cautious Fletcher and his second, equally cautious Spruance, we would have lost our carriers in a rush to follow-up the first day's action and have run into Yamamoto's battleships in the night, as was the Japanese plan.

Fletcher was not a "dashing warrior", but a task force commander who had to husband the few American resources. He "dashed" to the attack at Tulagi and again to successfully cut off the Japanese invasion of Port Moresby at the Coral Sea. He raced to Midway and victory. And he was the first to attack in the Eastern Solomons against an enemy double his size. Fletcher commented after his success in forcing the enemy to retreat at that battle, that "he would receive a message of congratulations on a great victory from Adm Nimitz and a complaint from Adm King that he should have followed up the air battle with a destroyer attack -- and that both would be right."

> Note that the Japanese force escorting their troop convoy included three battleships : *Mutsu, Hiel,* and *Kirishima* ; 13 heavy cruisers ; 3 light cruisers ; and 31 destroyers; plus submarines and the two remaining fleet carriers, *Shokaku* and *Zuikaku,* plus another, light carrier, *Jintsu.* Admiral King had never fought in a war; he had captained a reserve destroyer after the Occupation of Vera Cruz (where Fletcher won a Medal of Honor) and was a flag assistant in WWI (where Fletcher commanded destroyers in Europe to earn a Navy Cross). King represented Washington-style dreams, not reality.

If Fletcher had stayed in the South Pacific instead of being moved to the North Pacific, he might have sunk another six enemy carriers and ended the war in another eight months.

Nobody had ever fought a carrier war before and everybody could see other ways "they" would have fought -- after the battles had concluded. But Fletcher was there and he won all of his battles -- which no other admiral had done to that time. What more can we ask?

Lastly, Fletcher had retired; most of his papers of the early months of WWII went down with *Yorktown* and he chose not to reconstruct his papers from the Pentagon basement and sit for Samuel Eliot Morison's definitive *History of U.S. Naval Operations in World War II*. Therefore, wartime activities were presented without Fletcher's point of view and favored other admirals who personally told their stories. He was presented as a hesitant, if proper officer. In point of fact, **Fletcher was our combat commander who saved the Pacific in the first months of the Pacific War** while other navies – U.S. Battle Fleet, British Far East Fleet, Netherlands East Indies fleet, U.S. Asiatic Fleet, and British Indian Ocean fleet – collapsed. He gave America time to build up to a superior fighting force. The time he bought also allowed the U.S. to tackle "Germany first", when the American people were most concerned about Japan.

Overview   23

He sank six enemy carriers with a loss of but two. He performed a nearly perfect balance of caution and aggression. In the two months after Fletcher left, his successors lost two carriers, yet sank no more enemy carriers for almost two years.

Questions and Answers

This is a compilation of Questions and Answers, often from Wikipedia which gets changed by random people over time who sometimes represent gross misunderstandings of World War II, and we need a permanent record here.

Example:

1. AT GUADALCANAL, THE NAVY RAN AWAY WITHOUT UNLOADING ALL THE SUPPLIES AND MEN.

Hint, *I learned recently on Wikipedia to avoid certain emotional triggers, even if true.* I had said, "the Marines on Guadalcanal may not have had all the typewriters and cheese unloaded that they would have liked." There was OUTRAGE! This morning, I had to point out that :

The Marines had personally "combat loaded" their ships (because the New Zealand stevedores would only work a 5-day week, and that was Wednesday to Sunday to get double time for weekends.) Combat loading is prioritizing -- putting ammo and rations on top, with typewriters and cheese below. However, the ships that sailed directly from the States to Guadalcanal were not combat loaded.

2. A MARINE RANTED THAT NOT EVEN ALL THE TROOPS WERE LANDED.

Sorry, but these were the floating reserve, and not needed during the uncontested landing on Guadalcanal, and were tasked to occupy Ndeni Island on the third day. This was canceled when that island was occupied by seaplane tender USS *McFarland* (AVD-14).

3. ANOTHER SAID THE NAVY FIGHTS SAFELY FROM AFAR, NOT LIKE BRAVE MARINES, AND DID NOT EVEN LOSE A DESTROYER.
To which I had to respond:

Gross error -- Guadalcanal was a battle at sea. More sailors were killed in the first two days defending that beachhead than were Marines in the entire six-month land battle. While 1,592 U.S. Marines and Army soldiers died on land, 48 warships went down – half ours, half enemy : 3 carriers, 2 battleships, 12 cruisers, 25 destroyers, 6 subs. The total lives lost at sea remains unknown (to me). We do know 1,270 Allied sailors died in only the first of six great sea battles at Guadalcanal.

4. AND ANOTHER SAID FLETCHER WAS NOT THE ONLY "FIGHTING" ADMIRAL.

"There were other fighting admirals during these painful times – Adm Pye had to refloat his battleships ; VAdm Brown had to retreat from Rabaul, attacked New Guinea with Fletcher, no kills ; RAdm Fitch, Fletcher's air adviser at the Coral Sea, lost *Lexington* ; VAdm Halsey was in hospital during this period ; RAdm Kinkaid, new boy under Fletcher in Solomons, no kills. ; RAdm Mitscher, sent ashore for poor handling of *Hornet,* no kills ; RAdm Murray, promoted, aggressive, lost *Hornet,* no kills ; RAdm Spruance, with Halsey's staff, was under Fletcher at Midway. (Note, the win belongs to the man in charge and giving the orders – It was Fletcher who sent an order to Spruance at 0607 : *"Proceed southwesterly and attack enemy carriers as soon as definitely located."*) ; RAdm Noyes, arrived and lost *Wasp* in ten weeks with no kills ; VAdm Fletcher, Coral Sea, Midway, and Eastern Solomons, lost *Yorktown,* while making six enemy carrier kills.

Would you say he did an admirable job? Perhaps one could say with a near-perfect balance of aggression and caution? Others performed well during this difficult period, but this is Fletcher's page, and he was the most successful." -- *I will hear about this paragraph.*

5. AND LASTLY, SOMEBODY OBJECTED TO MY USE OF THE WORDS "RAMPAGING ENEMY".

During the first months of WWII, when we were losing the war, the Japanese Combined Fleet had defeated the U.S. Battle Feet, British Far East Fleet, U.S. Asiatic Fleet, Netherlands East Indies Fleet, and the British Indian Ocean Fleet in attacks from Hawaii to India, from Alaska to Australia without loss of any ship to the strike fleet and none larger than a submarine or auxiliary destroyer in their whole Navy. What other words would be used other than to say this was a *rampaging enemy?* They were later defeated by overwhelming American industrial might ; we built 119 carriers during the war ; the Axis built ten. That story of the later victories can be read on other pages. But, with the remnants of a depression era fleet, Fletcher was the most successful of our Admirals in stopping them -- three times -- and always with a smaller force. These successes allowed the U.S. the time to mobilize.

> "The Japanese *rampaged* through the Pacific until they met Fletcher." -- del Castillo, *The Days of Fletcher.*

6. OH, YEA. SOME IDIOT CHALLENGED THAT WE WERE REALLY OUTNUMBERED IN THE PACIFIC.

The ship names on December 8th were :

Japanese :
Carriers (10) : Akagi , Kaga , Soryu , Hiryu , Shokaku, , Zuikaku , Hosho , Ryujo , Zuiho , Taiyo.
Battleships (10) : Kongo , Hiei , Kirishima , Haruna , Fuso , Yamashiro , Ise , Hyuga , Nagato , Mutsu.
Yamato, world's largest battleship, commissioned later in the month; that makes eleven.

American, British, Dutch, and Australian :
Carriers (3) : Lexington , Saratoga , Enterprise.
There are four USN carriers in the Atlantic War -- Ranger, Yorktown , Wasp , Hornet.
British HMS *Hermes* was in the Indian Ocean.
Battleships : U.S. –**none** ; R.N. – **(2)**
USS Colorado (BB-45) was in overhaul. There were eight in the Atlantic on "neutrality patrol" against the Nazi.
HMS **Prince of Wales** and battlecruiser HMS **Repulse** were sunk on December tenth ; that makes **none**.

7. ANOTHER ONE. A FELLOW COMPLAINED THAT FLETCHER VIOLATED ORDERS AND ONLY CAME WITHIN 260 MILES OF MIDWAY WHEN ORDERED TO 200 MILES.

Point Luck, the meeting point on June 2, was 325 miles NE of Midway. Fletcher moved the fleet to 260 miles north of Midway on June 3rd and to 200 miles for the dawn of June 4th. This plan was discussed back at Pearl before departing. Ships cannot sit at a point, they operated in a block cruising at about twice the speed of a submerged submarine. Ships try never to reverse course in case they run into just such a chasing submarine, but move laterally in a rough box. TF 16 operated, 15 miles south of TF 17. In fact Spruance was heading NW within the waiting area (away from Midway) when he received Fletcher's order to proceed southwesterly. The Japanese fleet was proceeding diagonally towards Midway and was a little south when first sighted. Fletcher moved his forces west and a bit south to intercept.

If one gets out a chart and plots latitude and longitudes, one finds interesting things. Spruance launched from 173 miles from Midway and Fletcher was further north. The whole battle took place about 200 miles north of Midway and the distance from Midway is not particularly relevant, rather ship to ship distance is. Virtually all drawings of the Midway operations use a square of equal degrees E-W and N-S. At that location on the globe, 20 deg N-S is equal to the distance of 15deg E-W. By showing on an equal scale graph, thus elongating N-S distances, it looks on paper as if the fight was much more on a diagonal then it really was, flights were essentially east-west. Texts then follow the wrong impression given by the drawings : "Nagumo changed course NE, (070 degrees)". We nautical types know that NE is 045deg, not 070 deg. He turned East and less than 2 points from due East, (using the 32 compass points that WWII sailors were taught) : or just over one point north, East by North (now 78.5 deg) and less than two points off East Northeast (now 66.5 deg). In fact, most authors use words taken from the graphs. Perhaps it is just a literary simplification to call things northeast and southwest to indicate just a little bit north or south of a due East-West line.

Hint: pay attention to the numbers, not the words.

8. WHICH ADMIRAL WON MIDWAY?

There were four American admirals involved with the Battle of Midway. Chester Nimitz was CINCPAC, Commander in Chief of the Pacific Fleet. He took the great risk of sending his whole fleet in an attempt to ambush the enemy. Frank Jack Fletcher was his commander of two task forces sent to wait and fight at Midway. Raymond Spruance headed one task force, TF 16, with two fleet carriers, *Enterprise* and *Hornet*. These had been under the command of Vice Admiral Halsey, who was hospitalized on his return from the Doolittle raid on Tokyo and a rush to the South Pacific. Spruance had been selected to become the next Chief of Staff for Nimitz. He was next in line for command and he needed combat experience to help him in the Staff job. Although he had no carrier experience, it was thought that Halsey's staff would serve him well. That didn't turn out to be so. The captain of *Hornet,* Marc Mitscher, had just been promoted to Rear Admiral. Fletcher wore two hats, one as commander of Task Force 17 with *Yorktown,* and second as Officer in Tactical Command of the combined Task Forces with Spruance reporting to Fletcher and with Mitscher reporting to Spruance. For completeness, there were admirals with each cruiser screen. RAdm Thomas C. Kinkaid with TF 16 and RAdm William W. Smith in TF 17.

9. BUT HISTORY BOOKS SAY FLETCHER TURNED OVER COMMAND TO SPRUANCE.

Spruance went on to become Chief of Staff of the Pacific fleet under Nimitz. He is said to have had great intellectual capacity and was needed in that capacity to plan the American return to the Pacific and then was given command of the Fifth Fleet to carry out that plan. His biographers went back to the first time he appeared on the world stage, which was at Midway and

Overview

self-servingly attributed the victory there to him. After all, his task force had Halsey's two carriers and Fletcher only had one. But it was the same one with which Fletcher had stopped the enemy at the Coral Sea and Fletcher was the combined task forces commander.

It was he who on hearing scouting reports of the sighting of two carriers and a battleship -- when four or five carriers were expected -- that ordered Spruance "to proceed southwesterly and attack the carriers as soon as the location is definitely known." It was Fletcher that held *Yorktown* in reserve to attack, with his one carrier, the other two or three enemy carriers expected – location unknown. Halsey's staff was excited to attack and they attacked prematurely without following Fletcher's experienced advice to know the location the enemy.[1]

Enterprise's bomb-laden planes flew to the general area then spent over an hour searching empty sea before finding the enemy. *Hornet's* two bomber squadrons never did locate them. The flights were disorganized and the torpedo bombers went in alone -- to the well known tragic results in which 37 of 41 were slaughtered without one hit upon the enemy. The full-scale dive bomber attack by *Enterprise* planes saw Bombing 6 attack *Akagi*, while Scouting 6 attacked *Kaga*. Meanwhile, during this hour (80 minutes) that
TF 16 planes were trying to find an enemy fleet in the open sea, Army and Marine planes from Midway attacked more than two enemy carriers, It is a scandal that land plane radio was not intercepted by the carriers. In a *use'em or lose'm* situation, Fletcher decided to commit half of his reserve to the known two targets. They flew directly to the enemy, arriving at the same time as the two previously lost *Enterprise* squadrons, attacked *Soryu,* and returned. Many *Hornet* and *Enterprise* planes ditched for lack of fuel. Pilots from a whole fighter squadron from *Hornet* went into the sea together. Unfortunately, Spruance had not the experience to pass judgment on the excitable planning of his inherited staff, and history suggests that only the strength of a Halsey could have controlled that bunch. *Hornet's* two bomber squadrons totally missed the battle. Because of this *Hiryu* was not attacked. She was ready with torpedo planes which followed the American squadrons back to their ships. *Yorktown* was spotted and attacked. *Yorktown* planes landed on the less then full decks of *Enterprise* and *Hornet.* Captain Buckmaster had *Yorktown* sound and carrying on flight operations by the time a second wave of Japanese bombers arrived that ended her operations for the day. But Fletcher as commander was searching for the fourth and fifth enemy carriers and had already sent out scouts and urged Midway to extend their search, too. *Hiryu* was found by his *Yorktown* scouts and ten bombers from *Enterprise* and fourteen bombers originally from *Yorktown* were launched and sank the fourth enemy carrier.

After a second attack, with *Yorktown* listing badly, abandon ship was ordered and Fletcher transferred to *Astoria* and became engaged in saving his flagship and its crew. At this point, *Enterprise* approached and Fletcher released TF 16 to continue the battle for the next day. It was dusk and lights had to be put on to land the returning bombers. Fletcher proceeded east and Spruance followed when the returning bombers were landed so as to prepare for his operations of tomorrow.

We can see that Spruance succeeded in attacking the Japanese as ordered, though not as successfully as if he had more closely followed Fletcher's instructions. In fact, this directly led to the enemy being able to launch the attack on *Yorktown*. Fletcher commanded the battle, assigning, finding and attacking in turn, until his flagship was badly damaged.

The next morning, Spruance, delayed by a submarine scare, returned to the mutual protection of Midway aircraft, assured himself that the invasion had been called off, and went in search of the retiring Japanese Combined Fleet. They were out of range and nothing of importance happened that day. The third day he attacked two damaged cruisers that had collided avoiding U.S. submarine *Tambor* (SS-198). One of the cruisers was sunk and Spruance withdrew his task force to refuel. The battle was over.

----

    (1) Mistakes made by Halsey's staff that Spruance had not the experience to recognize were wrong.

    • Point Option. Point Option is a moving path of where returning pilots can expect to find their home. *Enterprise* staff failed to consider the delay necessary to alter course into the wind to launch and recover aircraft and was not able to advance as the pilots had been told. This means TF 16 was further way than the returning planes could find without searching, many had not the fuel to land. The staff never realized their error, else they could have had a destroyer follow the path of Point Option to direct the returning planes home.

    • Rush to Attack. If Point Option had been correctly calculated, then the attack could not have been rushed because the return flight would have assured that neither fighters nor torpedo planes could be expected to have made the return.

Overview   29

- Forgetting about *Hornet*. *Enterprise* was Spruance's flagship for TF 16 that included *Hornet*. Twice the staff forgot to tell *Hornet* the plan and the time for launching planes. *Hornet* correctly had her squadrons ready and waiting. The first time without word, the pilots were returned to their ready room, then were launched with only 22 minutes notice and without the latest information. The second instance, that afternoon, having received no word, *Hornet* broke the spot to board the bombers that had had to divert to Midway when they ran out of fuel looking for the enemy that morning. They had to re-spot the planes and follow Enterprise planes (most were *Yorktown*'s) by 25 minutes, missing a coordinated attack.
- We do not fault Spruance for splitting the initial *Enterprise* air attack force. There was a foul-up aboard that delayed launching in the middle. Spruance had to either land and refuel those planes already in the air, or to send them ahead. He had no choice but to send them ahead. Murphy's Law.

10. BUT WASN'T FLETCHER PUTTING THE ROUGH PART OF THE BATTLE ON SPRUANCE AND HOLDING BACK FROM THE FRAY?

Look, Fletcher fought three of the five great carrier battles in all of history. Let's look at his performance in each. At the Coral Sea he gave up part of his own defenses by sending Crace's cruisers to stop the enemy invasion transports. Fletcher then turned on two enemy carriers and succeeded in turning back both the transports and the carriers, sinking one. At Midway, Fletcher sent Spruance with two carriers to attack two enemy carriers, Meanwhile, Fletcher reserved the task for himself to handle the two or three other carriers[2] expected with only his one carrier. He sank four carriers.

At Eastern Solomons, Fletcher attacked a force double his size and did not wait for reinforcements from his own units refueling nearby. He turned back a force twice his size and sank a carrier. In eight months he sank six enemy carriers and lost two. Wouldn't you say that was courageous? Fletcher reserved the biggest tasks for himself. If there was any shyness, it was on the part of the Japanese who were defeated or ran away from the smaller forces of Fletcher.

----

(2) There could have been as many as seven other enemy carriers out there at Midway . *Soryu* and *Hiryu* had not been sighted in the original PBY report. Only one of two carriers had succeeded in attacking Dutch Harbor, *Ryujo*, so the location of *Junyo* was unknown. *Zuikaku* had her flight crews replaced from the Coral Sea and showed up late at Midway and was sent to Alaska to ambush the U.S. force. *Shokaku* was badly damaged at the Coral Sea, but so had been *Yorktown* and she made it to the battle. *Zuiho* was with the transport convoy and approaching Alaska. *Hosho* was with Yamamoto's Main Body.
Was Fletcher brave or dumb? Actually he was following the highest traditions of duty to attack superior enemy forces, just as the torpedo plane pilots attacked against almost certain death.

BUT I KEEP READING THAT FLETCHER TURNED OVER COMMAND TO SPRUANCE.

Why does everybody say he turned over command of the battle to Spruance? It was AFTER the enemy was *destroyed!!!!!* that Fletcher *released* Spruance's TF 16 to continue the battle for the next day while Fletcher salvaged TF 17. On his own for the next two days, Spruance only sank one of two previously damaged cruisers. Fletcher had a **4 : 0** ballgame going at the time he released Spruance -- nobody attributes the game to the relief pitcher when the team is ahead by 4 to 0 when he takes over.

FLETCHER WAS WRITING THE BOOK ON CARRIER WARFARE.

At the Coral Sea, *Lexington* was damaged to an extent that should have been salvageable later in the war. Fletcher introduced flushing empty fuel lines with nitrogen to prevent explosions of "empty" lines. At Midway, his pilots were ordered with magnetic "vectors", whereas the rest of the Navy spent a lot of effort to teach sailors how to calculate "arrow" headings from true north. Pilots had magnetic compasses. Didn't it make sense to avoid making pilots do calculations while trying to navigate a fighter or bomber on a mission?

Peacetime doctrine launched bombers first so as to reach altitude while torpedo planes and fighters were launched and all proceeded together. Fletcher launched by flight speed -- first the slow torpedo planes, then the bombers that were faster even when climbing, and lastly the short-legged, fast fighters which allowed all to catch up to make a combined attack on the target.

Point Option was known to all who were concerned with returning aircraft. Fletcher practiced it when others had to learn it.

Fletcher wanted a flag officer on each carrier – ship's captains had enough on their minds without having to coordinate with the evolving battle.

He liked two-carrier task forces with each carrier operating separately in combat so as to be able to freely maneuver, but close enough to support each other with a fighter umbrella. At Midway and Guadalcanal, he further separated his 3rd carrier to similar effect and to make it less likely for attackers to find both task forces. It worked.

MYTHS OF MIDWAY

The idea of ambushing of the Japanese carriers was a spark of genius. It was also a spark of desperation. Given the conditions, it was the best that could be hoped for. We were losing the war. Washington was all over Nimitz. He had some secret information of Japanese plans, so he risked all. There was no legitimate expectation that the Japanese could not find out that two task forces were at sea -- only because the Japanese subs were late; their flying boats could not land at French Frigate Shoals; the Hawaiian spies; and *Tone*'s catapult malfunction all went awry. That is too thin to be in a realistic plan. We should have lost at Midway. Nimitz was either more of a genius than we can understand, or crazy/bedeviled by higher management to do something, ... anything -- like a man in debt seeking salvation in a casino. Later activity proves he was not crazy, but was surely lucky.

- The thought that Spruance was the victor because he was the admiral with the most carriers. He followed orders as part of his learning curve for higher command. He had Halsey's staff. He could never have been blamed for a defeat; that would fall on Nimitz and Fletcher, thus he is not blamed for errors, we blame Halsey's staff instead. He was just too inexperienced to recognize the many errors they made.

- The idea that Spruance had the foresight to turn away at night and brilliantly escaped Yamamoto's battleship trap. Carriers are unable to attack at night and therefore defenseless. There made no sense to advance in darkness when there were two known battleships with Nagumo and who knew how many surviving and other carriers? Nobody knew Yamamoto was coming with additional battleships and carriers. *It didn't matter.* The U.S. carriers could not risk confronting those battleships already known about. To advance into the dark would have been foolish and Spruance might have been inexperienced, but he was not a fool. He followed Fletcher away for the night, turned around at midnight intending to arrive back to the battle zone by first light.

- That the Japanese carriers were sunk while Spruance was operating independently. Fact, two of the Japanese carriers *sank* the next day. But they were attacked, hit, burned, and put out of commission the first day by Fletcher. Even famous historians make this careless error.

MORISON'S ERRORS

A good example for people to be able to understand historian Morison's errors is from an example by Morison himself. In Book III, Page 159, he humorously has a long paragraph about the Army Air Force claiming the victory at Midway "with hits on 3 carriers, 1 cruiser, 1 battleship, 1 destroyer and 1 large transport." This was printed in all the press and the public thought air power was grand against warships. The public had to be gradually disabused of this error. Horizontal bombers *never* hit, let alone sank, a moving warship. Likewise, Morison's venomous errors against Fletcher must be gradually corrected. This is a "Morison errors" paragraph that may well become its own web page; the final straw to me was him having Fitch aboard *Saratoga* during Midway.

Other topics -- Morison says Fletcher missed Wake because he stopped to refuel. In fact, he was ordered to wait in place for another carrier to join up with him and made a good use of this time by refueling. Then Morison said he should have done it the *previous* day when the weather was better ; how stupid, the wait order had not been issued then.

Fuel consumption errors: Morison doesn't seem to know that fuel usage goes up exponentially with speed -- he uses convoy consumption rates in his calculations whereas battle speeds use multiple times more fuel.

He suggests that a low fuel state can be allowed by destroyers. More stupidity. (German destroyers ran out of fuel at Norway and were helplessly sunk by the Royal Navy.)

Then he suggests that cruisers could refuel destroyers -- while within range of enemy bombers. Madness.

He was against Fletcher's rotation of refueling at Eastern Solomons. Yet, by keeping two carriers on station, Fletcher was able to repulse the enemy. If all three had refueled at the same time the enemy could have retaken an undefended Guadalcanal.

The Days of Fletcher has a full chapter of Morison errors.

Very bad, of course, is his publishing of the summary *Two Ocean War* in 1963, long after he knew Fletcher had been ordered to wait for *Lexington* (VAdm Brown) as reinforcement at Wake and by refueling Fletcher was making good use of waiting time. And he never did revise the text in the many reprints of the 15 volume set.

And the greatest sin is of allowing pique at Fletcher's reluctance to come out of retirement (his papers had been lost with Yorktown) to help write Morison's books is allowed to color his portrayal in those books and subsequent ones.

USS Panamint (AGC-13) Fletcher's last home at sea.

"Araby. Fletcher's home in Maryland, now a historic site for its colonial origins

Frank Jack Fletcher

From Wikipedia, the free encyclopedia -- 4 July 2008

Frank Jack Fletcher (April 29, 1885 – April 25, 1973) was an admiral in the United States Navy during World War II. Fletcher was the operational commander at the pivotal Battles of Coral Sea and of Midway. He was the nephew of Admiral Frank Friday Fletcher.

Frank Jack Fletcher
April 29, 1885(1885-04-29) – April 25, 1973 (aged 87)

Vice Admiral Frank Jack Fletcher, USN Photo aboard USS Saratoga, 17 September 1942. Official U.S. Navy Photograph

Place of birth	Marshalltown, Iowa
Place of death	Bethesda, Maryland
Allegiance	United States of America
Service/branch	United States Navy
Years of service	1906-1947
Rank	Admiral

Overview 35

Battles/wars	Veracruz (1914) World War I World War II –Battle of the Coral Sea –Battle of Midway –Guadalcanal campaign –Tulagi –Eastern Solomons
Awards	Medal of Honor Navy Cross Navy Distinguished Service Medal
Relations	Nephew of Frank Friday Fletcher

Contents

- 1 Early life and early Navy career
- 2 World War I and post-War period
- 3 Interwar service
- 4 World War II
    - 4.1 Wake Island — December 8 - December 23, 1941
    - 4.2 January - April 1942
    - 4.3 Coral Sea — May 4 - May 8, 1942
    - 4.4 Midway — June 4 - June 7, 1942
    - 4.5 Landing at Guadalcanal — August 7 - August 9, 1942
    - 4.6 East Solomons - August 24 - August 25, 1942
    - 4.7 Northern Command
- 5 Postwar and final days
- 6 Medal of Honor citation
- 7 Legacy
- 8 Notes
- 9 See also
- 10 External links

Early life and early Navy career

Fletcher was born in Marshalltown, Iowa on April 29, 1885. Appointed to the U.S. Naval Academy from his native state in 1902, he graduated from Annapolis on February 12, 1906, and commissioned an Ensign on February 13, 1908 following the required two years at sea.

The early years of his career were spent on the battleships *Rhode Island, Ohio,* and *Maine.* He also spent time on USS *Eagle* and USS *Franklin.* In November 1909 he was assigned to USS *Chauncey,* a unit of the Asiatic Torpedo Flotilla. He assumed command of USS *Dale* in April 1910 and March 1912 returned to *Chauncey* as Commanding Officer. Transferred to USS *Florida* in December 1912 he was aboard that battleship during the occupation of Vera

Cruz, Mexico, in April 1914. For distinguished conduct in battle at Vera Cruz, he was awarded the Medal of Honor (see citation below).

World War I and post-War period

Fletcher became Aide and Flag Lieutenant on the staff of the Commander in Chief, U.S/ Atlantic Fleet in July 1914. After a year at this post, he returned to the Naval Academy for duty in the Executive Department. Upon the outbreak of World War I, he served as Gunnery Officer of USS *Kearsarge* until September 1917, after which he assumed command of USS *Margaret*. He was assigned to USS *Allen* in February 1918 before taking command of USS *Benham* in May 1918. For distinguished service as Commanding Officer USS *Benham*, engaged in the important, exacting, and hazardous duty of patrolling European waters and protecting vitally important convoys, he was awarded the Navy Cross.

From October 1918 to February 1919 he assisted in fitting out USS *Crane* at San Francisco. He then became Commanding Officer of USS *Gridley* upon her commissioning. Returning to Washington, he was head of the Detail Section, Enlisted Personnel Division in the Bureau of Navigation from April 1919 until September 1922.

Interwar service

He returned to the Asiatic Station, having consecutive command of the USS *Whipple,* USS *Sacramento*. USS *Rainbow*, and Submarine Base, Cavite. He served at the Washington Navy Yards from March 1925 to 1927 ; became Executive Officer of USS *Colorado* ; and completed the Senior Course at the Naval War College, Newport in June 1929-30 and followed immediately by the Army War College in Washington, D.C. 1930-31 in preparation for strategic leadership responsibilities.

Fletcher became Chief of Staff to the Commander in Chief, U.S. Atlantic Fleet in August 1931. In the summer of 1933 he was transferred to the Office of the Chief of Naval Operations. Following this assignment he had duty from November 1933 to May 1936 as Aide to the Secretary of the Navy, the Honorable Claude A. Swanson. He assumed command of USS *New Mexico*, flagship of Battleship Division THREE in June 1936. In December 1937 he became a member of the Naval Examining Board, and became Assistant Chief of Bureau of Navigation in June 1938. Returning to the Pacific between September 1939 and December 1941 he became Commander Cruiser Division THREE; Commander Cruiser Division SIX; Commander Cruisers Scouting Force; and Commander Cruiser Division FOUR.

World War II

Wake Island — December 8 - December 23, 1941

Responding to reports from U.S. Marines on Wake Island of Japanese bombardment and a subsequent invasion attempt in the first week after Pearl Harbor, Fletcher was sent west with the carrier *Saratoga* (Task Force 11) to provide relief. He was one day away when plans were changed and ordered to wait for *Lexington* (Task Force 12, Vice Admiral Brown).[1] The next day the Japanese successfully invaded Wake Island. The task force was recalled by Admiral Pye, who was "keeping the seat warm" until Admiral Nimitz could arrive at Pearl and take over as Commander-in-Chief, Pacific Fleet.

Fletcher was one day away from engaging any enemy forces attacking Wake Island. He is often criticized for not arriving on station at Wake in time to defend the island. His desire to maintain "frequently refueling operations" for his destroyers to keep them ready for "high speed chase" is often cited as the critical delaying factor.[2] Both Fletcher and Pye have been criticized as exhibiting "poor seamanship and decision making".[3]

January - April 1942

On January 1, 1942, Rear Admiral Fletcher took command of Task Force 17 built around the carrier *USS Yorktown* (CV-5). He, a surface fleet admiral, was chosen over more senior officers to lead a carrier task force. He learned air operations on the job while escorting troops to the South Pacific. He was junior TF commander under tutelage of the experts: Vice Admiral William Halsey on raids in the Gilbert Islands in February; Vice Admiral Wilson Brown attacking the enemy landings on New Guinea in March; and had aviation expert Rear Admiral Aubrey Fitch with him during the first battle at theCoral Sea.

Coral Sea — May 4 - May 8, 1942

In May 1942, he commanded the task forces during the Battle of the Coral Sea. This battle is famous as the first carrier-on-carrier battle fought between fleets that never came within sight of each other.

Fletcher with *Yorktown,* Task Force 17, had been patrolling the Coral Sea and rendezvoused with Rear Admiral Aubrey Fitch with USS Lexington, Task Force 11, and a tanker group. Fletcher finished refueling first and headed West. On hearing the enemy was occupying Tulagi, TF 17 attacked the landing beaches sinking several small ships before rejoining *Lexington* and an Australian cruiser force under Rear Admiral John Gregory Crace on May 5.

The next day intelligence reported a Japanese invasion task force headed for Port Moresby, Papua New Guinea, and a Carrier Strike force was in the area, The morning of May 7 Fletcher sent the Australian cruisers to stop the transports while he sought the carriers. But first he sank the Japanese aircraft carrier *Shōhō,* escorting the enemy troopships, -- "Scratch one flat top." Meanwhile, Japanese carrier planes of Rear Admiral Chuichi Hara found the American tanker, USS *Neosho* (AO-23*)*, and sank it with its destroyer, *Sims*.

The Japanese launched a dusk patrol of 27 bombers that found nothing, but was detected on radar and attacked by Wildcats that shot down nine ; eleven more splashed while attempting night landing.

On 8 May, at first light, "round three opened." Fletcher launched seventy-five aircraft, Hara sixty-nine. Fitch had greater experience in handling air operations, and Fletcher had him direct that function, as he was to do again later with Noyes at Guadalcanal. *Shokaku* was hit, but not damaged below waterline; it slunk away. *Zuikaku* had earlier dodged under a squall. The Japanese attack put two torpedoes into *Lexington*, which was abandoned that evening. *Yorktown* was hit near her island, but survived. Hara failed to use *Zuikaku* to achieve victory and withdrew. The invasion fleet, now without air cover, also withdrew, thereby halting the Port Moresby invasion. Fletcher had achieved the objective of the mission at the cost a carrier, tanker, and destroyer. In addition, his Wildcats had beaten Japanese air groups, 52 to 35, and had damaged *Shokaku*. Neither Japanese carrier would be able to join the fight at Midway the following month.

This was the first time the Imperial Japanese Navy had been stopped. In their rampage across the Pacific from Pearl Harbor, East Indies, Australia, Ceylon; they defeated the British, Dutch, and Asiatic Fleets; and had not lost a fleet ship larger than minesweepers and submarines — until they met Fletcher.

Midway — June 4 - June 7, 1942

In June 1942, he was the Officer in Tactical Command at the Battle of Midway with two task forces, his usual TF 17 with quickly repaired *Yorktown*, plus TF 16 with USS *Enterprise* and USS *Hornet*. Vice Admiral William Halsey normally commanded this task force, but became ill and was replaced by Rear Admiral Raymond Spruance. When aircraft from four Japanese carriers attacked Midway Island, the three U.S. carriers, warned by broken Japanese codes and waiting in ambush, attacked and sank three enemy carriers – *Akagi, Kaga, Soryu*. *Enterprise* and *Hornet* lost seventy aircraft. Return attack damaged *Yorktown*. Fletcher's scouts found the fourth carrier and *Enterprise* with *Yorktown* planes then sank *Hiryu*. At dusk, Fletcher released Spruance to continue fighting with TF 16 the next day. During the next two days, Spruance found two damaged cruisers and sank one. The enemy transport and battle fleets got away. A Japanese submarine, *I-168*, found crippled *Yorktown* and sank her and an adjacent destroyer, USS *Hammann*. Japan had had seven large carriers (six at Pearl Harbor and one new construction) – four were sunk at Midway. This did not win the war, but evened the odds between Japanese and American fleet carriers.

Landing at Guadalcanal — August 7 - August 9, 1942

As the U.S. took the offensive in August 1942, Vice Admiral Fletcher commanded the invasion of Tulagi and Guadalcanal. Close air support was provided at Tulagi. The invasion of Guadalcanal was uncontested, Fletcher withdrew his carriers from dangerous waters when they were no longer needed. Rear Admiral Richmond K. Turner's offloading of supplies did not go as well as expected, he did not tell Fletcher, and then had to withdraw the transports after

Fletcher left. The Marines refer to this as the 'Navy Bugout', but the 17,000 Marines were in little danger from a construction battalion. The few U.S. carriers could not be risked against multi-engine, land-based, torpedo bombers, when they were needed for combat against carriers. He chose to withdraw on the third morning to prepare for the inevitable Japanese counterattack.

A separate incident must be mentioned :

The Battle of Savo Island - 9 August 1942.

Allied warships under Rear Admiral Victor Crutchley, RN, screening the transports were surprised at midnight and defeated in 32 minutes by a Japanese force of seven cruisers and one destroyer, commanded by Japanese Vice Admiral Gunichi Mikawa. One Australian and three U.S. heavy cruisers were sunk, and one other U.S. cruiser and two destroyers were damaged in this lopsided Japanese victory. However, as Crutchley notes, the transports were not touched. Fletcher is sometimes criticized because his carriers were at the far end of their nightly withdrawal, steaming back for the morning, yet too far to away to seek revenge.

Eastern Solomons - August 24 - August 25, 1942

Two weeks later, Fletcher used the carriers he had saved when he fought a superior Japanese fleet intent on counter-invasion in the carrier aircraft Battle of the Eastern Solomons. He started the engagement and sank his sixth carrier, *Ryujo*. The ensuing battle was essentially a giant aerial dogfight interspersed with ship-borne antiaircraft fire. The U.S. lost 20 planes, the Japanese lost 70. *Enterprise* took a couple of bombs and *Chitose* was nearly sunk, but survived. The enemy withdrew without landing troops on Guadalcanal. They had to resort to the "Tokyo Express" : overnight delivery of a few hundred troops and supplies by destroyers. Fletcher, as always, was second guessed by non-combatants, and was criticized by Admiral Earnest King, in Washington, for not pursuing the Combined Fleet as it withdrew. This criticism may have affected the decision to not return Fletcher to his command after his flagship, the carrier *Saratoga* (CV-3), was torpedoed and damaged by a Japanese submarine on August 31, 1942. Fletcher himself was slightly injured in the attack on *Saratoga*, suffering a gash to his head and was given his first leave after eight months of continuous combat. His successors lost two carriers in two months for no victories, such that in the first week of November there were *no* active U.S. carriers in the Pacific.

Northern Command

In November 1942, he became Commander, Thirteenth Naval District and Commander, Northwestern Sea Frontier to calm the public fear of invasion from the north. A year later, he was placed in charge of the whole Northern Pacific area, holding that position until after the end of World War II, when his forces occupied northern Japan.

Postwar and final days

Vice Admiral Fletcher was appointed to the Navy's General Board in 1946 and retired as Chairman of that governing board in May 1947 with the rank of full Admiral. He enjoyed life on his county estate, Araby, in Maryland.

Many of Fletcher's papers were lost in combat, he declined to reconstruct his papers from Pentagon archives and sit with Morison who was writing the naval history of World War II, and in return received no consideration by Morison, an attitude picked up by later authors.

Admiral Frank Jack Fletcher died on April 25, 1973, four days before his 88th birthday at the Bethesda Naval Hospital in Bethesda, Maryland. He is buried in Arlington National Cemetery.

Medal of Honor
Citation:

For distinguished conduct in battle, engagements of Vera Cruz, 21 and 22 April 1914. Under fire, Lt. Fletcher was eminent and conspicuous in performance of his duties. He was in charge of the *Esperanze* and succeeded in getting on board over 350 refugees, many of them after the conflict had commenced. Although the ship was under fire, being struck more than 30 times, he succeeded in getting all the refugees placed in safety. Lt. Fletcher was later placed in charge of the train conveying refugees under a flag of truce. This was hazardous duty, as it was believed that the track was mined, and a small error in dealing with the Mexican guard of soldiers might readily have caused a conflict, such a conflict at one time being narrowly averted. It was greatly due to his efforts in establishing friendly relations with the Mexican soldiers that so many refugees succeeded in reaching Vera Cruz from the interior.

Legacy

During the first months of WWII, when the Allies were losing the war, the Japanese Combined Fleet had defeated the U.S. Battle Feet, U.S. Asiatic Fleet, British Far East Fleet, and Netherlands East Indies Fleet in attacks from Hawaii to India, Alaska to Australia without loss of any ship larger than a submarine or auxiliary destroyer. Until they met Fletcher. He was never able to advance upon the enemy, but he was able to stop them -- three times, always with a smaller force. These successes allowed the U.S. to mobilize and to build 119 carriers during the war, while the axis built ten. That story of later victory can be read on

Overview *41*

other pages. But, with the remnants of a depression era fleet, Fletcher was the most successful of our Admirals in stopping a rampaging enemy. With a near perfect balance of aggression and caution, he sank six enemy carriers with the loss of two. Fletcher must be remembered as the task force commander that held the line to allow time for later, grand victories.

USS *Fletcher* (DD-992) is named in honor of Admiral Frank Jack Fletcher.

Notes

1. Morison, Samuel E., *Supplement and General Index*, History of United States Naval Operations in World War II, Volume 15, Cumulative Errata
2. John Costello, *The Pacific War: 1941-1945*, Harper Perennial, December 1, 1982.
3. Morison, Samuel E., Admiral USN. *The Rising Sun in the Pacific,* 1931 - April 1942, History of United States Naval Operations in World War II, Volume 3.

United States Navy Portal

List of Medal of Honor recipients

External links

- Frank Jack Fletcher at *Find A Grave*. Retrieved on 2007-10-23.
- Layton, Edwin T. (with Roger Pineau and John Costello) (1985). *And I Was There: Pearl Harbor and Midway — Breaking the Secrets*.
- Admiral Frank Jack Fletcher, USN, (1885-1973). *Online Library of Selected Images: People - United States*. Naval Historical Center, Department of the Navy.
- Frank Jack Fletcher, Admiral, United States Navy. ArlingtonCemetery.net (26 March 2006).
- "http://en.wikipedia.org/wiki/Frank_Jack_Fletcher"
- Lundstrom, John (2006). *Black Shoe Carrier Admiral: Frank Jack Fletcher at Coral Sea, Midway & Guadalcanal*. Annapolis: Naval Institute Press. ISBN 1-59114-475-2.
- Lundstrom, John B. (2005 (New edition). *First Team And the Guadalcanal Campaign: Naval Fighter Combat from August to November 1942*. Annapolis: Naval Institute Press. ISBN 1-59114-472-8.
- Regan, Stephen D. (1994). *In Bitter Tempest, the Biography of Admiral Frank Jack Fletcher*. Iowa State University Press. ISBN 0-8136-0778-6.

- Morison, Samuel E., Admiral USN. *The Rising Sun in the Pacific, 1931 - April 1942, History of United States Naval Operations in World War II, Volume 3.*
- A Critical Revisit to the Battle of Midway. George J. Walsh, Lieutenant Commander. USNR (22 December 2006).

USS *Fletcher* (DD-992)
named for Admiral Frank Jack Fletcher

MARSHALL TIMES

VOL. 3, NO. 12   DEC. 1994

STORMY SEAS...

...the story of famous
Marshalltown Admiral
Frank Jack Fletcher

Marshalltown's Frank Jack Fletcher, Allied fleet commander in the Pacific Ocean Theatre of World War II, receives the Navy's Distinguished Service Medal from Chester Nimitz, commander in chief of the Pacific Fleet in 1942. The taciturn Fletcher has become a historical enigma over the past five decades. To find out why, turn to our cover story beginning on page 12.

Stormy Seas...

⑫ MARSHALLTIMES DECEMBER 1994

Frank Jack's Final Farewell...

Marshalltown's Admiral Fletcher Is Buried Next To The Grave Of President Kennedy

MARSHALLTIMES DECEMBER 1994 13

The Story Behind The Rise And Fall Of World War II's Top Navy Fleet Commander

Four white horses led Admiral Frank Jack Fletcher's casket up a winding hill in Arlington National Cemetery on a cloudy April afternoon in 1973. The funeral cortege came to a stop near the grave of John F. Kennedy, buried in Arlington less than ten years earlier.

And as Marshalltown's most famous military commander was laid to rest adjacent to the Kennedy monument, dozens of young visitors standing at the JFK gravesite turned their attention to the ceremonies only a few feet away. Just who was Frank Jack Fletcher, many of them undoubtedly wondered. And why is he being buried with full honors next to the burial site of the martyred President?

The baby-boomers at Kennedy's grave weren't alone. By the time that Admiral Fletcher died just four days shy of his 88th birthday, he had been out of the public eye for almost four decades. Few people under fifty even

MARSHALLTIMES feature by Iowa Journalist and Historian Paul "Biff" Dysart

remembered the reticent gardener, who lived quietly with his wife of 56 years on their enshrouded estate in suburban Maryland.

But the older folks remembered Frank Jack Fletcher. They remembered the days when they would sit by their living room console radios, straining to hear the voice of Walter Winchell above the crackle of the airwaves. And during the spring and summer of 1942, the name of Admiral Fletcher was heard more than any other, save Franklin Delano Roosevelt himself.

Everyone in America knew Frank Jack Fletcher back then...

To understand Frank Jack Fletcher, you must first understand late 19th century Marshalltown. It was a stratified society, where the sons of the privileged few went on to college, and money and social position were handed down, generation to generation, by the descendants of the Founding Families.

This, then, was the Aura of Privilege that Frank Jack Fletcher was born into during the spring of 1885. His grandfather was famous Marshalltown pioneer George Glick, whom Fletcher biographer Stephen D. Regan calls an "unabashed adventurer, promoter, backslapper, entrepreneur and go-getter". His father was prominent Marshalltown businessman and mover and shaker Thomas Fletcher. And his uncle was nationally-known Navy Admiral Frank Friday Fletcher.

To say the Glick and Fletcher families prospered in Marshalltown would be an understatement indeed: George Glick started the ball rolling by becoming a Democrat right before the election of President James Buchanan in 1856, just in time to reap the coveted position of Postmaster for the fledgling little prairie village of Marshalltown.

Glick then purchased a drug store, and began to run the Post Office out of the back of his store. It was there, according to Regan's biography "In Bitter Tempest", that Glick and "two of his cronies" planned for the land acquisition and platting of Riverside Cemetery.

Of course, it was also in Glick's drug store that the eager entrepreneur plotted to "divert" city funds to a Chicago-based railroad in an attempt to entice the line to make Marshalltown the rail center of Iowa. "The frontier courts, limited but not inept, overruled the promoters," Regan noted. "The town was forbidden to give municipal funds as outright gifts."

Early Influences

George Glick (above), Frank Jack Fletcher's grandfather, and Thomas Fletcher, his father

Overview

Frank Jack Fletcher Came From One Of Marshalltown's Most Prominent Families

Glick, undaunted, immediately proclaimed a citywide holiday and Marshalltown's businesses all shut down for a day, so every businessman could help Glick shake down the entire town for "voluntary" contributions for the railroad.

And yes, the railroad yards came soon after Glick's successful "town holiday"...

A year after the railroad first came through Marshalltown George Glick was elected township trustee. He immediately engineered the incorporation of Marshalltown, and by 1868 was representing the Fourth Ward on Marshalltown's first City Council.

Alice Glick Fletcher

By 1875 he was also president of the bank that he and several compatriots had launched, and he was instrumental in persuading the Iowa legislature to locate the Veterans Home in Marshalltown. Later in life he helped start Bankers Life Insurance Company in Des Moines.

His daughter Alice was the youngest in a family of four brothers, and she became a teacher of "underprivileged children" after her graduation from Marshalltown High School. Several years later she married young banker Tom Fletcher, who worked for her father, and the family settled into the Glick family home (now the Marshall County Historical Society) at the corner of Second Avenue and Church Street.

MARSHALLTIMES Vol 3 #12
Published monthly by Gramam/Dysart and Associates
P.O. Box 1513, Marshalltown, Iowa 50158
Editor : Paul "Biff" Dysart
Used by permission of the publisher of the
Times-Republican, *August 24, 2010.*

He Stars in Football at MHS...

The Glicks And The Fletchers Secure A Naval Academy Spot For The 1902 Grad

It was there that young Frank Jack Fletcher grew up as the 19th century wound down, and he lived an active childhood in the shadows of his family's constant and expansive activities. His father became president of Marshalltown's First National Bank when his grandfather moved to Des Moines to launch Bankers Life, and his mother Alice Glick Fletcher became the leading lady of Marshalltown society by the turn-of-the-century.

Alice helped form the Hawthorne Club and later helped chart the Iowa Branch of the Federation of Women's Clubs. When her husband Tom joined with Frank Letts to found Marshalltown's public library, Alice served on the library's first board of trustees.

While Tom Fletcher wheeled and dealed as Marshalltown entered the 20th century, and while young Frank Jack Fletcher was an accomplished MHS scholar and football star, Alice was a whirlwind of activity. She helped form the Elmwood Country Club, was a charter member of the Twentieth Century Club, and together with her friend Mary Julia Letts helped their husbands form the well-known Letts-Fletcher Wholesale Grocery Company.

And she made sure that the new elementary school going up southwest of Marshalltown's expanding business district was named for her father!

Frank Jack Fletcher concentrated on English, Greek and Roman history at MHS, along with various levels of science and mathematics, preparing thoroughly for the formal education that his family would expect of him later. But some of young Fletcher's favorite childhood moments had come during the Marshalltown visits of his famous uncle Frank Friday Fletcher, an American Admiral by the time the Spanish-American War broke out in 1898 following the explosion of the USS Maine in Havana Harbor.

"Friday's letters and visits must have been tremendous occasions, with revelations about the people he had met and adventures he had experienced in Australia, South America and Asia, especially attending the treaty negotiations that opened Korea to international trade," Regan wrote. "He was on the battleship Maine in Cuba, but he had transferred just prior to its explosion."

So not just any prestigious university would be enough for Frank Jack Fletcher when he graduated from MHS in the spring of 1902. He wanted to go to the United States Naval Academy in Annapolis, Maryland.

The Glick-Fletcher family influence network went to work that summer to insure that young Frank Jack would receive Iowa's only appointment to the Academy. Uncle Friday Fletcher waved his rank in front of Iowa's United States Senators, and grandfather George Glick invited the entire Iowa Congressional Delegation to dinner at his Des Moines mansion,

General of the Army Hap Arnold presents Fletcher with an Army Distinguished Service Medal.

Overview 49

MARSHALLTIMES   DECEMBER 1994   19

Fletcher Promoted To Admiral As The War Clouds Gather...

Receives Medal Of Honor, Navy Cross

where the subject of the state's appointment to the Naval Academy just happened to come up.

And by the autumn of 1902, young Frank Jack Fletcher was duly enrolled in the prestigious academy, on the fast-track to his moment of destiny some forty years later...

Marshalltown's Admiral Fletcher arrives on a Navy inspection tour...

Fletcher was a good student at Annapolis, but Regan notes that football, his first love in Marshalltown, was all but ignored at the academy. He applied himself academically, especially during his final two years, and finished 30th in a class of 116 of the best and brightest young men in America in 1906.

He was immediately assigned to the battleship USS Rhode Island, and became a commissioned officer in 1908. For the next decade he served in various capacities all over the world, and as World War I approached he was named commander of the destroyer USS Benham, relieving his Annapolis schoolmate Comdr. William "Bull" Halsey, Jr. Fletcher spent the summer of 1918 chasing submarines and protecting convoys, and he quickly caught the eye of officials in the Woodrow Wilson administration.

Later that year the Marshalltown man was awarded the Navy Cross for distinguished service, and the dashing young officer and his new bride (Kansas City socialite Martha Richards) returned to Washington, D.C. amid glowing tributes. Regan notes that his Washington years in the 1920's were important for his career, as he made the acquaintance of prominent military and political figures who he could count on to assist him later in life.

"Fletcher had the insight in the 1920's to see that the Asiatic Fleet, long thought to be the dead-end of Naval Careers, could be used for advancement," Regan wrote. "He adroitly foresaw that China and Japan were going to be two dominant forces, and that a thorough understanding of the Pacific and the nations involved in the Western Pacific area would be beneficial someday."

By the 1930's Frank Jack Fletcher and his charming wife Martha were among the toast of the Washington social set, and when the Admiral was stateside they entertained lavishly at their colonial estate named Araby, only a short hop from the nation's capital. And when he was ordered back on active duty during a crisis, he used each incident to add another feather to his professional cap: He helped lead an invasion of the Philippines to quash a guerrilla network, he quelled an insurrection in Haiti and he stormed into Vera Cruz harbor during another international incident as America's

Navy truly began to rule the waves of the Western Hemisphere.

By the late 1930's he had earned the prestigious Navy Medal of Honor to go with his Navy Cross, and in 1940 he was promoted to Rear Admiral and given command of the USS Minneapolis.

And as Frank Jack Fletcher weighed anchor in Pearl Harbor, Hawaii on Dec. 5, 1941, he was reaching the pinnacle of his Navy career. And as war clouds gathered across the Pacific Ocean, few American Admirals had Fletcher's combat experience, acute understanding of Japan or more knowledge of surface warfare. As biographer Regan put it many years later: "His time had come."

Fletcher's USS Minneapolis left Pearl Harbor on Saturday, Dec. 6, with a fleet of cruisers, and he heard about the Sunday morning attack by the

Scores Triumphs At Coral Sea, Midway

Fletcher Takes Command Of Pacific Fleet

Japanese as he was preparing for breakfast southwest of Oahu. He returned to Pearl Harbor immediately.

The American command became increasingly concerned about holding tiny Wake Island after the attack on Pearl Harbor, and Fletcher was selected to lead the Wake Island relief efforts. Delays and refueling problems, however, kept Fletcher's command from reaching Wake in time to defend it, and the strategic island was taken by the Japanese.

The Wake Island failure marked the first of repeated broadsides aimed at Fletcher from Admiral Ernest King and others in the Department of the Navy, but the Marshalltown man was still named commander of the Pacific Task Force on New Year's Day, 1942. Fletcher led the task force into the first major naval battle of World War II four months later at Coral Sea, and although both the Americans and Japanese forces suffered heavy losses, Fletcher finally had produced the major victory that the American public had been hungering for all through the winter and the spring.

"Overall the Coral Sea battle was a very important one for the American cause," Regan wrote. "Though outnumbered and outgunned, Fletcher won. It stopped Japanese expansion in the Pacific, it seriously crippled Japan's carrier depth for the next battle, and it was the turning point that moved momentum in America's favor."

Frank Jack Fletcher had little time to rest on the laurels of Coral Sea, however. A Japanese force was bearing down on Midway, another crucial American military base, and Fletcher's task force steamed away to help defend the island. In one of the major naval and air battles of the war, Fletcher's task force and the American command at Midway repelled the invasion, but again at great cost. Fletcher was once again criticized for unnecessary delays and lack of decisiveness.

Admiral Fletcher (left) and his dog Peg treat Soviet Naval officers to an early 3-D film showing on board the USS Minneapolis.

But many historians have since come to Fletcher's defense, albeit after the fact. "Midway was the great benchmark of the war because it ended the defensive period of the U.S. Navy," Regan wrote. "With his victories, culminating at Midway, Fletcher gave (the Allies) their first real opportunity to try some hard-hitting attacks."

Back on the homefront, Central Iowans listened intently to KFJB and WHO for every little detail as hometown hero Frank Jack Fletcher led the Americans through the Solomon Islands campaign, through Guadalcanal and into the famous Battle of Savo Island on August 9. America's industrial power was slowly turning the tide of the war in the Pacific, and Fletcher could take heart that his continual early skirmishing had at least won a war of attrition during the first few months in that theater.

But his enemies in Washington were continually back-stabbing the Rear Admiral, some claiming that Fletcher's inherent conservatism had kept the Allied forces from pursuing the enemy when the Japanese were clearly on the defensive. Of course, Fletcher had never enjoyed a numerical superiority in either ships, guns or manpower during the first six months of 1942.

That's why many Americans were outraged when Fletcher was removed from his command by the Navy

Frank Jack Accepts Japanese Surrender At Matsu Bay In '45

...but doesn't return to Marshalltown

Department and dispatched to the 13th (Northwest) Naval District in Seattle to serve out the remainder of the war. He had made a fatal enemy in the high-ranking but desk-bound Admiral King, who had the ear of Franklin Roosevelt, and not even Fletcher's friends in high places could prevent King's sudden transfer of America's most experienced fleet admiral.

After the nuclear attack on Hiroshima and Nagasaki brought a sudden end to the war, Fletcher again regained a degree of national prominence. He accepted the surrender of the Japanese Imperial Northern Fleet in Mutso Bay in 1945, and quietly retired from active duty soon after that as he reached his 60th year. He had come a long way from his studies at Abbott Elementary School on the corner of Linn Street and Fourth Avenue a full half-century earlier...

Frank Jack Fletcher never returned to Marshalltown after the war. His family was largely gone by the 1930's (he had no children and only one nephew) and he and his devoted wife Martha retired to their beautiful estate at Araby. Martha still hosted bridge games and dinner parties for old friends and they both spent hours tending their spacious flower gardens, but by the 1960's the Admiral was approaching 80 and began to suffer through a series of health-related problems.

By the early 1970's he was spending much of his time at the Bethesda Naval Hospital, where he passed away as the Cherry Blossoms bloomed once again along the Potomac. And his widow Martha draped his command flag over the casket as his funeral entourage began its long procession to Arlington Cemetery on that grey day in 1973.

Frank Jack Fletcher might have been a Marshalltown boy. But in the end, Admiral Frank Jack was just a Navy Man....

Admiral Frank Jack Fletcher accepts the surrender of the Japanese Imperial Northern Fleet at Matsu Bay in 1945.

REMEMBER
Back Issues of
MARSHALLTIMES
magazine are available at Marshalltown's public library

Admiral Frank Friday Fletcher
The first "Admiral Fletcher from Iowa"

Born on November 23, 1855 at Oskaloosa, Iowa, graduated from the United States Naval Academy in 1875. While assigned to the Bureau of Ordnance he developed the Fletcher breech mechanism that increased the speed of rapid-fire guns. Fletcher was assigned to the battleship USS *Maine* when it blow up in Havana Harbor (he was absent that night) triggering the Spanish-American War in 1898.

As commander of U.S. Naval Forces on the East Coast of Mexico in 1914, he occupied the city of Vera Cruz and was awarded the Medal of Honor.

He became commander of the Atlantic Fleet, receiving promotion to Full Admiral and during World War I served on the Navy General Board.

He died November 28, 1928, and was buried in Arlington National Cemetery.

He was the uncle of World War II Admiral Frank Jack Fletcher.

The USS *Fletcher* (DD-445), lead ship of a class of 175 WWII destroyers, was named for Admiral Frank Friday Fletcher.

Medal of Honor

Citation:

FLETCHER, FRANK FRIDAY, Rear Admiral, U.S. Navy, "For distinguished conduct in battle, engagements of Vera Cruz, 21 and 22 April 1914. Under fire, Rear Admiral Fletcher was eminent and conspicuous in the performance of his duties; was senior officer present at Vera Cruz, and the landing and the operations of the landing force were carried out under his orders and directions."

The Fletcher Class of Destroyer

USS Fletcher (DD-445)

USS *Fletcher* (DD-445) commissioned 30 June 1942 and arrived at Noumea, New Caledonia, 5 October 1942, from the east coast, and at once began escort and patrol duty in the Guadalcanal operation, bombarding Lunga Point 30 October. *Fletcher* participated in the naval Battle of Guadalcanal, the Battle of Tassafaronga, sank Japanese submarine *RO-102*, supported the landings on the Russell Islands and New Georgia, invasion of the Gilbert Islands, made a strike on Kwajalein, Taroa, and Wotje, supported the Humboldt Bay landings, the invasions of Noemfoor, Sansapor, and Morotai, covered the occupation of the Philippines from Leyte, Luzon, and Manila Bay where she took a hit that killed eight crewmen. USS *Fletcher* received 15 battle stars for World War II service and five for Korean War service.

Specifications of the Fletcher Class

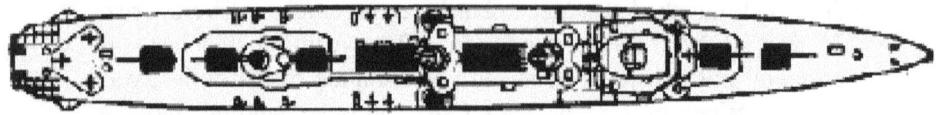

Displacement.	2,100 tons	Armament.
Length.	376' 3"	5 - 5"/38' dual purpose guns
Beam.	39' 8"	2 x 5- 21" torpedo tubes
Draft.	13'	6 depth charge launchers.
		2 depth charge tracks.
Speed.	36 knots	
Complement.	273	
Power.	twin screws	
	60,000 h.p.	

About 175 Fletcher class destroyers were commissioned during World War II from 1942 to 1944. Eighteen were lost to enemy action, all in the Pacific, eight from kamikazes off Okinawa in 1945. Most were sent to the reserve fleet after the war, superseded by the heavier Sumner class of 1944 with 55 ships, and the Gearing class of 1945 with 80 ships. Fletcher class ships were reclassified as type DDE in 1949. USS *Fletcher* herself was modernized and fought in Korea and Vietnam before decommissioning in 1969.

PART 2. Till They Met Fletcher

The U.S.S Yorktown And the Japanese Carrier Fleet
RENDEZVOUS AT MIDWAY
by Pat Frank and Joseph D. Harrington

FOREWORD
Admiral Frank Jack Fletcher (retired)

The war was only a few weeks old when I took my flag and my staff aboard U.S.S. *Yorktown* at San Diego, and our situation in the Pacific was indeed unhappy. Hong Kong and Wake had fallen. The two British battleships on which we had counted so much for stopping a Japanese southward advance had been sunk off Malaya. Allied strength in the western Pacific was but a handful of cruisers, some destroyers, and a few submarines, no match at all for the onrushing Japanese fleet.

The Philippines had been invaded at many points. General Douglas MacArthur had declared Manila an open city and was moving his forces into the Bataan Peninsula. There, as the war plans ordered, he was to fight as long as he could, then retreat into Fortress Corregidor and wait for us in California and Hawaii to come to his aid.

We could give no help, of course. The battleships that were to crash through the central Pacific and combine with the British and Dutch forces against the Japanese lay in the mud of Pearl Harbor. The Japanese warlords, on the last day of 1941, were free to strike wherever it pleased them to do so.

Few persons recognized then, or remember now, how grand the Japanese strategy was and how close it came to being realized. Japan struck east in early December, immobilizing the main strength of the U.S. fleet with a surprise attack on Pearl Harbor. She was now striking south, to control or isolate Australia, the only logical place from which a great counterattack could be launched against her. She was also getting ready to strike west, across India, and into the Near East, joining up with German forces under General Rommel in Africa.

With Japanese victories, the Axis Powers would have held the great Eurasian landmass, with all its wealth and resources, the world island. America would have been isolated and dealt with at leisure.

In the Pacific Ocean, at the beginning of 1942, the United States had no means whatever of thwarting Japanese plans except for small task forces built around four aircraft carriers -- *Enterprise, Saratoga, Lexington,* and *Yorktown*. A fifth carrier, Hornet, would soon be on its way from the Atlantic. So it was that the carrier task force was thrust into war.

Our aircraft carriers had to perform two tasks : protect what holdings we still had, and keep the Japanese fleet off balance and dispersed by widely scattered, sudden attacks on enemy holdings. We had to keep this up until replacement and supplementary warships could be constructed, until America's industrial might could make itself felt. And Lord help us if the Japanese fleet ever assembled against us, outnumbered as we were.

A look at the globe will awe anyone who considers just how much of the world the Rome-Berlin-Tokyo Axis would have controlled had its master plan been successful. Hitler's troops at that time, remember, were enjoying success everywhere. They seemed well on their way to complete victory in Europe and in North Africa.

I had the privilege of commanding American sailors in two great carrier battles of the Pacific -- Coral Sea and Midway. The first battle saved Australia and marked a new experience for the Japanese fleet, retreat. The second battle broke the back of Japan's naval air arm, with four of her best aircraft carriers going down and hundreds of her best pilots lost.

In both battles, I fought from the bridge of U.S.S. *Yorktown*, a fine ship, with whose officers and men it was an honor to serve. The full contribution of Yorktown men to America's success in the Pacific has never been revealed until now. Security precautions kept it secret at the time it happened, and later happenings in the war, when we were obviously on the road to victory, overshadowed it. After the war, when all the facts became available and were cleared for release, they were not reported fully. Certain myths were repeated so often that they became accepted truths.

The authors of this book decided to tell *Yorktown's* story. They gathered and documented all the facts obtainable concerning her, then assembled the personal experiences of men who sailed in Yorktown, so that her story could be told in their words. When I was first approached for an interview, I was impressed with the wealth of information, some of it new even to me, that these gentlemen possessed. I was also impressed with their meticulous method of cross-checking every bit of information. I remain impressed with the final result of their work.

U.S.S. Yorktown was in the thin line of aircraft carriers which were all we had to deter the Japanese with until our forces were built up. She helped keep the enemy spread out. When he finally did concentrate his forces for attack, she and her men helped meet and defeat them.

Once the Japanese Navy lost four aircraft carriers at Midway, it lost its momentum and never recovered it. Australia was never threatened again. Nor was Hawaii. America was able to divert men and resources to Africa, England, and Europe, to stop Hitler, and to send others to Burma and India to assist in stopping the Japanese there. Victory became a matter of time.

Aircraft carriers like *Yorktown* added a new dimension to naval warfare. They guarded supply lines, hunted down enemy submarines, protected convoys and even their own escorting capital ships, beat off enemy air counterattacks, They softened up beaches for invasion, and rained destruction on enemy cities. They became an integral part, indeed the core, of our naval power. They remain so today. As I write this, they are off Vietnam, demonstrating their unique ability to apply as much or as little force as is necessary in backing up the foreign policy of the United States.

What carriers can do we first learned in ships like U.S.S. *Yorktown*, during the tough, trying days of early 1942. The battle lessons of the Coral Sea and Midway were applied with ever-increasing effectiveness later -- with such effectiveness, in fact, that we were able to reduce our pilot training program radically in 1944. Our pilots were so well trained and superior to the enemy by that time that their losses were far below combat expectations.

Most *Yorktown* men lived through the Coral Sea and Midway. Badly outnumbered in the war's early days, they fought bravely and survived. They then trained and led others in the ways of victory. I am pleased to know that, at last, their story is being told in detail, just as it happened.

FRANK JACK FLETCHER
U.S. Navy Admiral (Retired)

Araby, La Plata, Maryland, 1966

眞珠灣攻撃の機動部隊航跡圖
TRACK OF CARRIER TASK FORCE FOR PEARL HARBOR ATTACK
26 NOVEMBER–23 DECEMBER 1941
ALEUTIANS
註
Note
Hours and dates are Tokyo time.
機動部隊
Carrier Task Force
0600 26 Nov
HITOKAPPU
4 Dec
日本
JAPAN
TOKYO
11 Dec
Submarine sighted
7 Dec
16 Dec
9 Dec
0700 7 Dec
1300 7 Dec
0130 8 Dec
23 Dec
To Wake
20 Dec
MIDWAY IS.
WAKE I.
U.S. Pacific Fleet
HAWAIIAN IS

Pearl Harbor Anchorage
7:55 A.M. Dec. 7, 1941
Monaghan
Farragut
Dale
Aylwin
Henley
Patterson
Ralph Talbot
East
Loch
Selfridge
Case
Tucker
Reid
Conyngham
Whitney
Blue
Phoenix
McGrew
Point
Aiea
Bay
Phelps
MacDonough
Worden
Dewey
Hull
Dobbin
Middle
Loch
Pearl
City
Detroit
Solace
Allen
Chew
Ramsay
Gamble
Montgomery
Trever
Breese
Zane
Perry
Wasmuth
Raleigh
Utah
Tangier
Medusa
Curtiss
Mokunui Is.
Mokuiki Is.
Arizona
Nevada
Vestal
Tennessee
West Virginia
Ford Island
Naval Air Station
Utility Plane
Hangars
Oklahoma
Maryland
Neosho
Harbor
California
Patrol Plane
Hangars
Kuahua
Argonne
Sacramento
Ramapo
Pelias
Avocet
Helena
Oglala
New Orleans
San Francisco
St. Louis
Honolulu
CINCPAC
H.Q.
Waipio
Peninsula
Pearl
Dredge
Cachalot
Swan
Rigel
4 Subs
2 Tenders
Castor
Sumner
Pennsylvania
Bagley
Shaw
(in floating dry dock)
6 Destroyers
1 Minesweeper
Oil
Storage
Helm
Downes
Cassin
Officer
Club
District H.Q.
Hospital
Point
Naval
Hospital
Navy
Yard
Hickam Field
0
.5
1 mi
N
©1997 MAGELLAN Geographix(805) 685-3100 www.maps.com

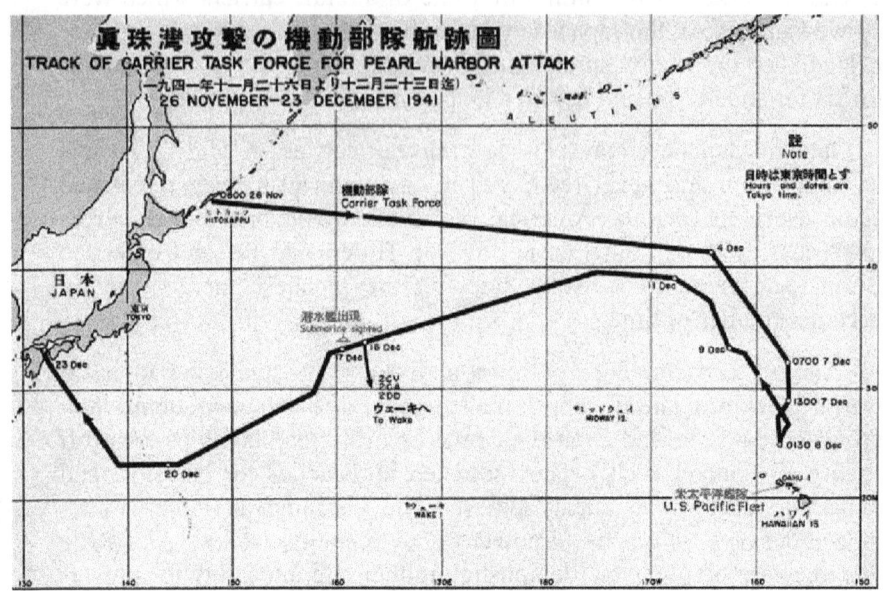

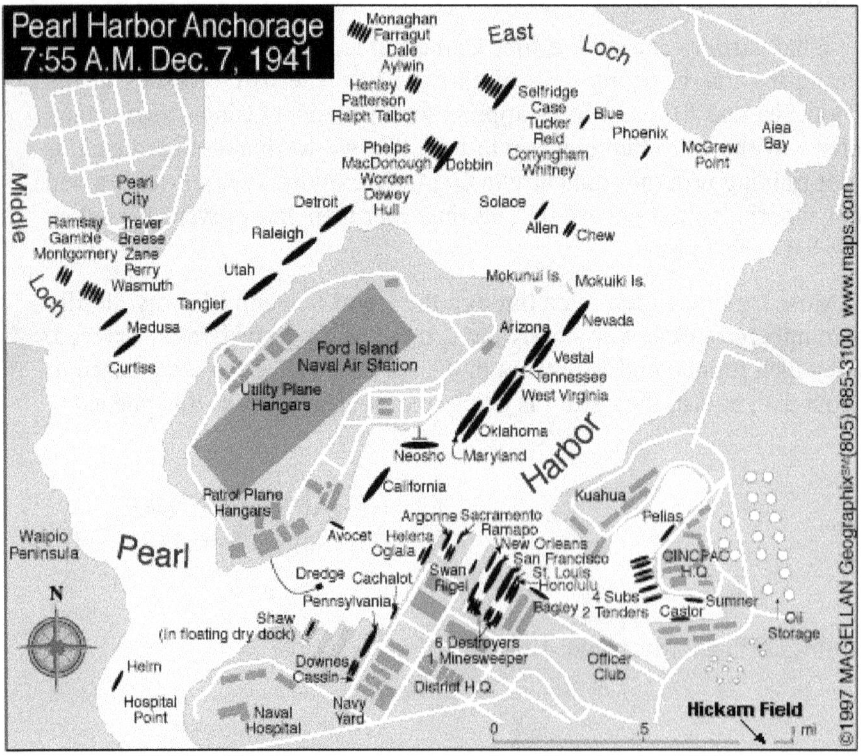

Why Did Japan Attack Pearl Harbor?

The attack on Pearl Harbor was part of a broad plan for Japan to take the leadership of Asia. They were not seeking war with the USA, but were trying to prevent the U.S. from warring with Japan.
* Japan entered World War I on the Allies side with the goal to take possession of the German colony of Kiaochow in China.
* Japanese troops made up the majority of Allied troops sent into Siberia, after Bolshevik Russia withdrew from the war, to keep supplies from being sent to Germany.
* Japan emerged in a strong economic, trade, and political position.
* In the Paris peace negotiations, Japan, having fought on the Allied side (and lost 301 men), was granted mandate over the German Pacific Islands north of the equator. Australia got those to the south.
* The U.S. convinced the Allies, England and Japan, to reduce naval levels to that of the depleted nations. Although Japan agreed to a smaller ratio of naval arms by the Naval Treaty, the number was greater than the Pacific half of the American allowance.
* England was pressured to give up her close relationship with Japan who was considered a rival by the United States. This removed all European influence over Japanese policy.
* Japan contributed to international trade with China by policing the railroads and eventually took over Manchuria as the puppet state of Manchukuo before the other world powers concluded that Japan had overstepped the bounds.

The end of World War I is said to have been just a lull in the fighting.
> *"The Peace Treaty of Versailles: This is not Peace.*
> *It is an Armistice for twenty years."* -- Foch, Allied Supreme Commander.

Nobody won that war, All European participants lost a generation of young men. Thirteen million died. Just as no one won the war, so too, no one won the peace. Except, perhaps, Japan. During the roaring 1920s, a fitting Japanese national policy was established.

1926 "Japan's divine mission is to rule the world."

1927 The Tanaka Memorial
"In order to conquer the world, we must first conquer China."
The Chinese government was backward and split among warlords and a Communist movement.

1937
- July 7. Start of Sino Japanese War – staged attack on Marco Polo bridge as an excuse to attack China proper.
  In First Sino-Japanese War of 1894-95, 42 years earlier, China was defeated and forced to cede Taiwan and Korea to Japan.
- Sep 25. Battle of Pinghsingkuan Pass. Chinese communists ambush Japanese.
- Oct 5. Emperor Hirohito removes constraints of international law on the treatment of Chinese prisoners.
- Nov 9. Japanese troops take Shanghai.
- Dec 8. Inner Mongolia is absorbed into the puppet state of Manchukuo.
- Dec 12. American river gunboat *Panay* sunk by Japanese aircraft while escorting evacuees from Nanking.
- Dec 13. Japanese sack Nanking - 300,000 murdered. Shocks the world.

1938
- July 29. Battle of Lake Khasan - Japan attacks Russia at Nomonhan on China/Russia border in Manchuria, north of Korea.
- August. Chiang Kai-shek withdraws his government to Chungking from The Japanese invasion.
- Aug 19. USSR repels the Japanese army near Vladivostok.
- October. The Japanese Imperial Army largely overruns Canton.
- December. President Roosevelt agrees to loan $25 million to Chiang Kai-shek. Throughout the war, Chiang is more interested in fighting internal communists than Japanese and must be bribed to do so.

1939.
- May 11. Battle of Khalkhyn Gol. Another Japanese thrust into Outer Mongolia is repelled by USSR.
- Jun 14. Tientsin Incident : Japanese blockade the British concession.
- Aug 20. Russian armored forces defeat Japanese in border war over Inner Mongolia.
- Sept 3. **War declared in Europe.**

1940
- May 7. Pacific fleet ordered to Pearl Harbor as a warning to Japan.
- July 5. U.S. bans strategic materials to Japan.
- Sep 27. Japan, German and Italy *Tripartite Pact* -- ten year military agreement - forms *Axis*.

1941

Jan 20. Japan orders cultural attaches in U.S. to establish intelligence gathering networks.

April 7 . Quarter of Pacific Fleet ordered to Atlantic against Nazis.

June 24. Japan pressures Vichy for IndoChina, to stop supplies to China.

July 29. Japanese start build-up of invasion troops from IndoChina.

Oct 13. Emperor expresses, "*Nothing ventured, nothing gained.*"

Oct 16. Japanese cabinet resigns ; new appointed by General Tojo.

Oct 16. First fleet caution sent to U.S. fleet : "Intelligence suggests Japan might attack Russia or British and Dutch Colonies in the East Indies . "

Nov 16. Japanese strike fleet to Kiriles to continue training.

Nov 18. Five Japanese mother subs, each with midget sub, depart Kure for Pearl Harbor.

Nov 24. CNO to U.S. Fleet Commanders : "... attack on Philippines or Guam is a possibility."

Nov 26. A Japanese strike fleet sorties towards Pearl Harbor

Nov 27. "War Warning": . . . an amphibious expedition against either the Philippines, Tai or Kra Peninsula or possibly Borneo."

Nov 30. Seaplane from Japanese submarine *I-10* reconnoiters Fiji.

Nov 30. Emperor orders PM Tojo to proceed with the attacks.

Dec 3 . Japanese invasion fleet departs Hainan for Thailand.

Dec 4 . Schedule of Pearl Harbor attack is transmitted to the sub fleet.

Dec 6 . Twenty-seven invasion transports depart Formosa for Philippines. 400 pilots briefed.

Dec 7 . Japanese invasion of Malaya.

Dec 7 . Japanese raid on Pearl Harbor.

Dec 7 . Air raids on Singapore, Guam, Wake, Philippines.

Dec 7 . Two Japanese destroyers shell Midway Island.

Dec 8 . Imperial Rescript declares a state of war between the Japanese Empire and the United States and United Kingdom.

Dec 8 . U.S. declares a state of war to exist with the Empire of Japan.

Dec 9 . Japanese planes sink HMS *Repulse* and *Prince of Wales* off Malaya.

When Did the Shooting Start?

The first shell fell on Kota Bharu, Malaya, 85 minutes before the first bomb fell on Ford Island, Hawaii. It was while 185 strike aircraft from six aircraft carriers were en-route on the 1-1/2 hour flight to Pearl Harbor, and while five miniature submarines were already trying to enter the inner harbor, an act of war.

Because USS *Ward* (DD-139) fired on and sank a mini-sub over an hour before the first bomb fell on Ford Island, a spin-master might say that the USA fired the first shot and started the war with Japan.

The more significant point is that the authorization to expand their war with China to the entire Pacific was given by the Emperor several days before these events. That decision sent two prepared Japanese fleets, one east, one west, and several ready task forces, to start an attack at 1800 GMT, 7Dec1941. It was all part of a master plan – a submarine tracked an American ship for several hours and exactly at 8 a.m. *I-26* surfaced and sank the *Cynthia Olson* 1,000 miles northeast of Diamond Head.

The reason Japan attacked Malaya was 1) to shut down the supply line to China through Burma, and 2) to protect their flank when they attacked the East Indies for the resources they needed for pursuit of the conquest of China. In their strategic planning, the Japanese felt a need to occupy the Philippines which sat on the supply-line between the Dutch Islands and the Homeland. The Philippine Islands were an American territory, therefore the Japanese planners saw a need to protect their back by eliminating the U.S. Navy at Pearl Harbor. This brings into strong light why the attack on Pearl was so casual and was not pressed with vigor -- there was no intent of conquest, just to give a hard smack to the Americans to keep them away during Japan's real task of taking Dutch oil and other strategic resources.

The mistake the Japanese planners made was to assume the U.S. would not make a big deal out of the swipe at Hawaii or in losing the Philippines, because to respond to Japan would activate a much more serious response from their treaty partners -- the victorious German-Italian Axis. The British and Dutch were already engaged in losing a war in Europe. Even Japan's traditional enemy, Russia, was being defeated. Surely the U.S. would not be so foolish as to enter into a war that they were certain to lose, too.

All of the goals of the Japanese strategy were accomplished in short order -- the supply line to China from Burma was cut ; the oil, rubber, and tin were acquired ; a defensive perimeter was established along the international dateline on the Pacific side ; and the British were pushed across the Indian Ocean to Africa thereby opening a potential Japanese tie to the Axis in the Middle East. However, the U.S. put up with 6-months of defeat while Admirals Nimitz and Fletcher held the line so that the American people and industry could mobilize -- and the rest is history.

These are the young men who rose to fight the attack on Pearl Harbor.

Ships at Pearl Harbor During the Attack
Starting at the harbor entrance at the west and going approximately clockwise.
Outside Harbor Entrance
Antares (AKS-3) - stores ship entering harbor followed by mini-sub.
Condor (AMc-14) - coastal minesweeper on patrol sights mini-sub at 03:57.
Ward (DD-139) - destroyer, on patrol, sinks mini-submarine at 06:40.
Hospital Point
Helm (DD-388) - destroyer, en transit East Lock to West Lock for deperming.
Rail (AM-26) ; Bobolink (AM-20) ; Vireo (AM-52) ; Turkey (AM-13) - auxiliary minesweepers
West Loch
Pyro (AE-1) - explosives transport
Middle Loch
Ramsay (DM-16) ; Gamble (DM-15) ; Montgomery (DM-17) - light minelayers
Trever (DMS-16) ; Breese (DM-18) ; Zane (DMS-14) ; Perry (DMS-17) ; Wasmuth (DMS-15) - a minelayer and fast minesweepers
Madusa (AR-1) - repair ship
Curtiss (AV-4) – large seaplane tender
NW side of Ford Island
Tangier (AV-8) - seaplane tender
Utah (AG-31) - gunnery training ship. Ex. battleship
Raleigh (CL-7) - light cruiser, flag, DesFlot1
Detroit (CL-8) - light cruiser, flag, DesFlot2
East Loch
Monaghan (DD-354) - ready destroyer ; Farragut (DD-348) ; Dale (DD-353) ; Aylwin (DD-355) - destroyers.
Henley (DD-391) ; Patterson (DD-392) ; Ralph Talbot (DD-390)
Dobbin (AD-3) - destroyer tender.
  Phelps (DD-360) [DL] ; MacDonough (DD-351) ; Worden (DD-352)
  Dewey (DD-349) ; Hull (DD-350)
Whitney (AD-4) - destroyer tender
  Selfridge (DD-357) [DL] ; Case (DD-370) ; Tucker (DD-374)
  Reid (DD-369) ; Conyngham (DD-371)
Blue (DD-387) - destroyer
Phoenix (CL-46) - light cruiser
Solace (AH-5) - hospital ship
Allen (DD-66) ; Chew (DD-106) - old destroyers

Key : **bold = sunk** ;   *italic description = damaged in raid*

Battleship Row

Inboard	Outboard
Nevada (BB-36) -- battleship	
Arizona (BB-39) – battleship	*Vestal (AR-4) -- repair ship*
Tennessee (BB-43) – battleship	**West Virginia (BB-48) -- battleship**
Maryland (BB-46) -- battleship	**Oklahoma (BB-37) -- battleship**
Neosho (AO-23) -- oil tanker	
California (BB-44) – battleship	
Avocet (AVP-4) -- small seaplane tender	

Navy Yard   Going approximately from west to east.
  Shaw (DD-373) - destroyer
  Downes (DD-375) , Cassin (DD-372)
  Pennsylvania (BB-38) - battleship
  Cachalot (SS-170) - submarine
  Oglala (CM-4) - fleet minelayer
  Helena (CL-50) - light cruiser
  Argonne (AG-31) - HQ ship
  Sacramento (PG-13) ; Mugford (DD-389) , Jarvis (DD-393)
  Swan (AVP-7) - small seaplane tender
  Ramapo (AO-12) - oil tanker
  Rigel (AR-11) - repair ship
  Tracy (DM-19) , Preble (DM-20) , Schley (DD-103) Cummings (DD-365)[DL]
  Pruitt (DM-22) , Sicard (DM-21) ; Grebe (AM-43) ; Ontario (AT-13) tug
  New Orleans (CA-32) - heavy cruiser
  San Francisco (CA-38) - heavy cruiser
  St Louis (CL-49) - light cruiser
  Honolulu (CL-48) - light cruiser
  Bagley (DD-386) - destroyer

Submarine Base
  Castor (AKS-1) - stores ship
  Sumner (AG-32) - survey ship
  Hulbert (AVD-6) - seaplane tender destroyer
  Thornton (AVD-11) - seaplane tender destroyer
  Narwhal (SS-167) - submarine
  Dolphin (SS-169) - submarine
  Taulog (SS-199) - submarine
  Pelias (AS-14) – submarine tender

USCG Station Honolulu - under USN command
  Taney (WPG-37) – 327' cutter, 2- 5", 4- 3"
  Reliance (WSC-150) – 78', patrol craft, 1- 3"
  Tiger (WSC-152) – 78', patrol craft, 1- 3"

The Pacific Fleet NOT at Pearl Harbor
At Sea on Sunday, Dec 7, 1941

Training Exercises
TF 3 Johnson Island -- VAdm Brown
    Heavy Cruiser - *Indianapolis* (CA-35)
        Dorsey (DMS-1) - destroyer minesweepers
        Elliot (DMS-4)
        Southart (DMS-10)
        Long (DMS-12)
        Hopkins (DMS-13)

West of Hawaii -- RAdm Fletcher
    Heavy Cruiser - *Minneapolis* (CA-36)
        Lamberton (DMS-2) - destroyer minesweepers
        Boggs (DMS-3)
        Chandler (DMS-9)
        Hovey (DMS-11)
    McFarland (AVD-14) - seaplane tender destroyer - off Maui, HI

Submarine Force
        Tambor (SS-198) – Wake Island patrol, north.
        Thresher (SS-200) – returning from Midway Island.
        Triton (SS-201) – Wake Island patrol, south.
        Trout (SS-202) – Midway Island patrol
        Argonaut (SM-1/SS-166) – Midway Island patrol
        Plunger (SS-179) – Diamond Head
        Pollack (SS-180) – Mare Island to Pearl
        Pompano (SS-181) – Mare Island to Pearl
        Gudgeon (SS-211) – Lahaina Roads

TF 12 Pearl Harbor to Midway Island - RAdm Newton
    Fleet Carrier - *Lexington* (CV-2)
    Heavy Cruiser - *Chicago* (CA-29)
    Heavy Cruiser - *Portland* (CA-33)
    Heavy Cruiser - *Astoria* (CA-34)
    Destroyer - *Porter* (DD-356) [DL, 8- 5"]
    Destroyer - *Flusser* (DD-368)
    Destroyer - *Lamson* (DD-367)
    Destroyer - *Drayton* (DD-366)
    Destroyer - *Mahan* (DD-364)
    Diverted to search for the Japanese fleet.

TF 8 Returning from Wake Island - VAdm Halsey
    Fleet Carrier - *Enterprise* (CV-6)
    Heavy Cruiser - *Northampton* (CA-26) RAdm Spruance
    Heavy Cruiser - *Salt Lake City* (CA-25)
    Heavy Cruiser - *Chester* (CA-27)
    Destroyers - *Balch* (DD-363) [DL, 8- 5"] ,
        Dunlap (DD-384) ,
        Fanning (DD-385) ,
    Destroyers - *Benham* (DD-397) ,
        Ellet (DD-398) ,
        McCall (DD-400) ,
    Destroyers - *Maury* (DD-401) ,
        Gridley (DD-380) ,
        Craven (DD-382).
    Battleships had been left behind as unable to keep up.

Convoy to Manila -- passing thru the Phoenix Islands
    Heavy Cruiser - *Pensacola* (CA-24)
    MTB Tender - *Niagara* (PG-52/AGP-1)
    2 Navy transports - *Chaumont* (AP-5) ; *Republic* (AP-33)
    2 Army transports - USAT *Willard A. Holbrook* , USAT *Meigs*.
    3 Army leased transports - *Admiral Halstead, Coast Farmer,* and the
        Dutch motor vessel *Bloemfontein.*
    ? Hallmark, Gregg, a corvette?
    This convoy was redirected to Brisbane by way of Fiji.

Convoy from Manila – passing between Santa Cruz and Ellice Islands
    Heavy Cruiser - *Louisville* (CA-28)
    2 Army transports - *Hugh L. Scott* (ex-*President Pierce,* later AP-43) ;
    President Coolidge - leased space

In Transit
    Fleet Carrier - *Saratoga* (CV-3) – arriving San Diego
    Destroyer - *Litchfield* (DD-336) – escorting *Thresher* thru Hawaiian
        waters.
    Seaplane Tender - *Wright* (AV-1) – Midway Island to Pearl Harbor.
    Minesweeper - *Robin* (AM-3) – Johnston Island to Hawaii
    Submarine Tender - *Fulton* (AS-11)-- shakedown cruise, off Guatemala
    Transport - *William Ward Burrows* (AP-6) – Hawaii to Wake Island,
        ordered to Johnson Island.
    Transport - *Regulus* (AK-14) at Midway.

Till they met Fletcher   *71*

Kodiak Alaska
    Seaplane Tender - *Gillis* (AVD-12) at Yakutat Bay
Samoa
    Minesweeper - *Kingfisher* (AM-25) at Tutuila

USCG
    The Coast Guard had been transferred to the USN on 01Nov41 and there were a variety of cutters and patrol craft in coastal and territorial waters.

SOUTH EAST PACIFIC, South America
    Light Cruiser - *Richmond* (CL-9) – off Chile
    Light Cruiser - *Trenton* (CL-11) – moored Balboa, Panama
    Submarine - *Gar* (SS-206) – off Acapulco, Mexico

U.S. WEST COAST
San Diego, CA
    Light Cruiser - *Concord* (CL-10)
    3 Destroyers - WWI era 4-stackers
        Talbot (DD-114) , *Waters* (DD-115) , *Dent* (DD-116)
    4 Submarines – *S-18* , *S-23* , *S-34* , *S-35*

Mare Island Naval Shipyard, San Francisco, CA
    6 Destroyers - *Clark* (DD-361) [DL] , *Cushing* (DD-376),
        Perkins (DD-377) , *Smith* (DD-378),
        Preston (DD-379) , *Rathburne* (DD 113)
    5 Submarines – *S-27* , *S-28* ,
        Nautilus (SS-168) , *Cuttlefish* (SS-171) , *Tuna* (SS-203)
    3 oilers and other auxiliaries.

Bremerton, Puget Sound, WA
    Battleship - *Colorado* (BB-45) – in overhaul thru March 1942.
    Seaplane tender (destroyer) - *Williamson* (AVD-2) – in overhaul.

ASIATIC FLEET - Admiral Hart
Philippines
Manila, Mariveles Naval Base and Cavite Navy Yard.
   Much of the surface fleet had been dispersed from Cavite with the "war warning" of Nov 28.
4 Destroyers - *Pope* (DD-225) ; *Peary* (DD-226) ;
   Pillsbury (DD-227) ; *John D Ford* (DD-228)
27 Submarines - *S-37* , *S-38* , *S-40* , *S-41* , *Porpoise* , *Pike* , *Shark* (overhaul), *Tarpon* , *Perch* , *Pickerel* , *Permit* , *Salmon* , *Seal* , *Skipjack* , *Snapper,*, *Stingray* , *Sturgeon* , *Sargo* , *Saury* , *Spearfish* , *Sculpin* , *Sailfish*, *Swordfish* , *Seadragon* (overhaul) , *Sealion* (overhaul) , *Searaven* , *Seawolf.*
2 Gunboats - *Asheville* (PG-21) , *Tulsa* (PG-22). Three more *en route.*
1 Flag yacht/gunboat -- *Isabel* (PY-10).
6 Minesweepers - *Tanager* (AM-5) , *Finch* (AM-9) , *Quail* (AM-15) , *Lark* (AM-21) , *Whippoorwill* (AM-35) , *Bittern* (AM-36)
2 Sub Tenders - *Canopus* (AS-9) , *Holland* (AS-3) ;
   Otus (AS-20) had arrived 7Dec. for conversion.
2 Seaplane Tenders - *Langley* (AV-3) , *Childs* (AVD-1)
2 Tankers - *Pecos* (AO-6) , *Trinity* (AO-13)
1 Submarine Repair Ship - *Pigeon* (ASR-6)
3 Auxiliaries – *Napa* (ATO-32) , *Genessee* (ATO-55) at Cavite *;*
   Floating drydock *Dewey* (YFD-1) at Mariveles, Bataan
Standard Oil tanker *George G. Henry* --civilian charter

Elsewhere in the Philippines
   Heavy Cruiser - *Houston* (CA-30) -- Iloilo, Panya Island
   Light Cruiser - *Boise* (CL-47) -- Cebu
   Submarines - *S-36* , *S-39* – north and south of the Philippines.
   Seaplane Tender - *William B Preston* (AVD-20) -- Davao
   Small Seaplane Tender - *Heron* (AVP-2) -- Port Ciego, Balabac Island,
      Palawan
   Auxiliary cargo - *Gold Star* (AK-12) -- Malangas, Mindanao

Taiwan Straits Admiral Hart had recalled the units from China.
   Oahu (PR-6),*; Luzon* (PR-7) , *Mindanao* (PR-8) are en route.
   Met by *Pigeon* and *Finch*, arrived 8Dec.

Till they met Fletcher 73

Borneo
At Tarakan
    Light Cruiser - *Marblehead* (CL-12)
    5 Destroyers - *Paul Jones* (DD-230) ; *Stewart* (DD-224) ;
            Bulmer (DD-222*)* ; *Barker* (DD-213) ; *Parrott* (DD-218) .
At Balikpapan
    Destroyer Tender - *Black Hawk* (AD-9).

Enroute Balikpapan to Singapore - to support *HMS Prince of Wales.*
    4 Destroyers - *Alden* (DD-211) ; *John D Edwards* (DD-216) ;
        Whipple (DD-217) ; *Edsall* (DD-219) .

China, Yangtze River Patrol - three en route Manila.
    2 Gunboats left behind
        Wake (PR 3) -- stripped, used as radio station, captured.
        Tutuia (PR-4) blockaded up-river, given to China.
        Panay **(PR-5)** was sunk at Nanking by Japanese Navy 1937.
Civilian charter -- *President Harrison* -- run onto a reef to prevent
    capture. Salvaged as *Kachidoki Maru* and sunk Sept 1944 by collision.
Guam
    Minesweeper - *Penguin* (AM-33)
    Auxiliary - *Robert L. Barnes* (AG-27) oil storage
    Patrol Craft – *YP-16* and *YP-17 (*ex-USCG 75-footers)

ATLANTIC
A quarter of the Pacific Fleet had been ordered to the Atlantic 7 April 41 for Neutrality Patrol. These were returned to the Pacific within six months of Pearl Harbor, except 3 light cruisers (CL-40, CL-41, CL-42) which stayed with the Atlantic Fleet.

    Battleships
        Idaho (BB-42)   Iceland
        Mississippi (BB-41)   Iceland
        New Mexico (BB-40)   neutrality patrol
    Carrier - Yorktown (CV-5)   Norfolk, VA
    Light Cruisers
        Brooklyn (CL-40) , *Philadelphia* (CL-41),
        Savannah (CL-42) , Nashville (CL-43)
    Destroyers - DesRon 8 (5 DD) , DesRon 9 (6 DD)

United States Aircraft at Pearl Harbor
USAAC: Hawaiian Air Force (7th Air Force)
Hickam Air Base
33 Douglas B-18 Bolo 2-engine standard bomber, 1936
12 Boeing B-17D Flying Fortress 4-engine heavy bomber, 1939
13 Douglas A-20A Havoc 2-engine attack/light bomber, 1940
 2 Douglas C-33 (DC-2) 2-engine freighter

Wheeler Air Base
 6 Boeing P-26A Peashooter open cockpit pursuit, 1934
 6 Boeing P-26B Peashooter open cockpit pursuit, 1935
39 Curtiss P-36A Mohawk pursuit, 1937
87 Curtiss P-40B Tomahawk pursuit, 1940
11 Curtiss P-40C Kittyhawk pursuit, 1941
 3 Martin B-12, 2-engine medium bomber, 1934
 3 Grumman OAF-9 Goose observation amphibian
 2 Douglas BT-2 biplane basic trainer
 2 North American AT-6 Texan, advanced trainer
 1 Seversky AT-12A Guardsman advance trainer

Five Mohawks engaged enemy, flaming two, with the loss of one P-36.

Bellows Air Field
 6 North American O-47B observation plane
 2 Stinson O-49 Vigilant L-1 observation plane
 1 Martin B-12, 2-engine medium bomber, 1936.

Two P-40s on training at Bellows attempted to take off and were shot down before gaining altitude. Other pilots were killed on the ground.

Haleiwa Air Field
 2 Curtiss P-36 Mohawk pursuit
 8 Curtiss P-40 Kittyhawk pursuit

The only unattacked airfield launched P-40s and P-36 on at least eight sorties and is credited with downing nine Japanese dive bombers with the loss of a P-36 to friendly antiaircraft fire.

Till they met Fletcher

On the 7th of December 1941, there were 223 Army aircraft based in Hawaii.

Airplane	Total	Destroyed	Damaged	Combat Ready
B-12	3	0	3	0
B-17 D	12	4	4	4
B-18 A	33	12	10	11
A-20 A	12	2	5	5
P-40 B	87	37	25	25
P-40 C	22	5	15	2
P-36 A	40	4	20	16
P-26	14	0	0	14
Total	223	64	82	77

Twelve USAAC B-17 Flying Fortress bombers arrived during the attack: B-17C and B-17E. They were unarmed, stripped for the overseas journey to the Philippines. One was destroyed, three others badly damaged.

One B-24A (prototype Liberator) intended for a secret photo mission is seldom reported. It was destroyed in the attack.

Nine Curtiss A-12 Shrike single-engine, light bombers (1932) still remained in service in Hawaii : Hickam 7, Wheeler 2.

Marine Corp :

Ewa Marine Corp Air Station (pronounced e'vee)
11 Grumman F4F-3 Wildcat fighter, 1940
8 Vought SB2U-3 Vindicator scout/dive bomber, 1938
20 Douglas SBD-1 Dauntless scout/dive bomber, 1940
3 Douglas SBD-2 Dauntless scout/dive bomber, 1941
2 Grumman J2F-4 Duck utility floatplane, amphibian
1 Lockheed JO-2 Electra Junior, six seat transport
1 Sikorsky JRS-1 Baby Clipper, twin-engine, 18 pass. flying boat
2 Douglas R3D-2 (DC-5) 2-engine paratroop transport
1 North American SNJ-3 Texan advanced trainer

Midway NAS
18 Vought SB2U-3 Vindicator scout/bombers.

Wake NAS
12 Grumman F4F-3 Wildcat fighters, 1940, just delivered by *Enterprise*

U S Navy:
Naval Base Defense Air Force

Pearl Harbor (Ford Island) Naval Air Station
19 Grumman J2F Duck single engine utility amphibian
 9 Sikorsky JRS-1 Baby Clipper, 18 passenger amphibian flying boat
 2 Consolidated PBY-1 Catalina patrol bomber, flying boat, 1936

Puunene (Lahaina Roads) NAS
4 Beech JRB "18", 2-engine utility transport
4 Northrop BT-1 torpedo bomber, 1938
1 Grumman JRF Goose, 2-engine amphibian
1 Grumman J2F Duck, utility floatplane, amphibian

Kaneohe NAS
1 Vought OS2U Kingfisher amphibian
36 Consolidated PBY-5 Catalina patrol bomber, flying boat, amphibian, 1939

Pearl Harbor (Ford Island) NAS Patrol Base
15 Consolidated PBY-3 Catalina scout bomber, flying boat, 1937
18 Consolidated PBY-5 Catalina scout bomber, flying boat, amphibian, 1939

   The apparently large number of Catalinas were not available for scouting. Two squadrons had arrived 23 Nov 41 for training while in transit to outlying islands and new crews were being trained by the few experienced crews. A special target by the Japanese, only the three out on patrol survived.

USN Air Battle Force
Pearl Harbor (Ford Island) NAS
       (spares, repairs, replacement, and training)
8 Brewster F2A-3 Buffalo fighter, 1939
5 Grumman F4F-3 Wildcat fighter, 1940
5 Grumman F4F-3A Wildcat fighter, 1941
3 Douglas SBD-2 Dauntless scout/dive-bomber, 1941

   SBD scout bombers from *Enterprise* arrived during the attack. They had flown ahead of the returning task force that had just reinforced Wake Island. About five were lost to Japanese or ground fire and an equal number damaged.

Fleet floatplanes, normally 2 for a cruiser and 4 per BB, are not shown. These were typically Curtiss SOC Seagull or the newer Vought OS2U Kingfisher. Japanese photos of battleship row show planes aboard *CA, OK, WV*. There were possibly 20-30 floatplanes that are not included in land base counts.

USAAC Aircraft Types
Bombers

Martin B-12. Twin-engine, dual cockpit, medium bomber designed in 1932 and entered service in 1934 as B-10, 117 built for the U.S. and over 200 for export. First U.S. bomber to fight in Pacific was flown by Dutch. B-12 had larger engines with 32 built. Three remained at Wheeler Field.

Douglas B-18 Bolo. Twin-engine design was a militarized DC-2 airliner. Evaluated in Aug 1935, ordered into production, Jan 1936 for 133 aircraft with follow up orders in 1937-38 of 217 aircraft. This was the standard Air Corp bomber; 33 were with the 7th AF at Pearl Harbor. They were phased into reconnaissance work when replaced by B-17 for combat duty.

Boeing B-17 Flying Fortress. Four-engine heavy bomber designed in 1934, flew July 1935, delivery of 13 prototypes started Aug 1937. Delivery started for an order of 39 in June 1939. B-17B order for 38 flew July 1940 with 20 sent to the U.K. Twenty-some based at Pearl Harbor were ordered to the Philippines. Twelve B-17s remained based at Pearl Harbor for long-range sea defense. An estimated 200 were required to adequately protect the approaches to Hawaii.

A squadron of twelve unarmed B-17s were staging through Pearl Harbor for the Philippines and landing to refuel during the attack. B-17s were a key element in MacArthur's defense plans ; however, he did not allow the 35 that he had to make an immediate strike on Formosa and most units were destroyed on the ground by those aircraft attacking from Formosa. It was soon discovered that horizontal bombers could seldom hit a moving warship, but provided a long-range reconnaissance and attack role through the Pacific War. 13,000 were built, mostly for the European Theater.

Consolidated B-24 Liberator. This second generation heavy bomber first flew year-end 1939. Two dozen were in British service as LB-30 for long-range transport and anti-sub service by Coastal Command in 1941. One prototype, intended for a secret photo mission, was destroyed in the attack on Pearl Harbor. The production model, B-24D entered AAF service in June 1942 in North Africa. 19,000 were built in all variants including the Navy PB4Y and the single tailed PB4Y-2 Privateer where a crewman was the last man to die in the war.

B-18 Bolo

B-17 Flying Fortress

Attack Bomber
* **Douglas A-20 Havoc (Boston).** This twin-engine, light bomber was originally designed in 1936 and an upgraded version flew Aug 1939 with 100 ordered by the French for delivery by year-end where 12 attacked German columns when the war began in Europe. A squadron of twelve were at Hickam Air Base and survivors made the first bomber sortie in an attempt to find the Japanese fleet. Over 7,000 were built.

Fighters
* **Boeing P-26 Peashooter.** Designed in 1932 as the Army Air Corps' first all-metal fighter, open cockpit, fixed landing gear. 126 built. Out of service by 1940. Twenty-six AAC units sold to the Philippines in 1941.
* **Seversky P-35 Guardsman.** Delivery of 77 started July 1937. 48 were in the Philippines and only 8 remained at the end of the second day. This design led to the Republic P-43 Lancer with turbocharger with 272 ordered in 1940.
* **Curtiss P-36 Mohawk.** First modern design for an USAAC fighter ; all-metal, closed cockpit, retractable landing gear with a radial engine. Designed in 1935. 210 delivered in 1937. Popular export. Five of 39 in training unit flew during raid on Pearl Harbor, credited with two Zeros.
* **Bell P-39 Airacobra.** (P-400) designed for ground attack, it was a nimble low-level fighter and among the first to arrive at Guadalcanal. Became famous as a panzer killer in Russia. None at Pearl Harbor.
* **Curtiss P-40 Warhawk** series developed from the P-36 with a larger, in-line engine. 524 ordered in Aug 1939 with 200 delivered to the U.K. April 1940 and U.S. delivery in May. The U.K. ordered 930 more; China ordered 100 of which 90 served with the Flying Tigers. Sept 1940 the U.K. ordered 1,500. Superior to enemy fighters of 1941 except for the Zero and Messerschmitt. Pearl Harbor had a hundred. A total of 15,000 were built with service throughout the war.

Other Aircraft, In Transit :
2 Dutch PBY-5 Catalinas from the factory to Soerabaya, Java.
    Number Y68 is destroyed and Y69 is repaired.

Air Force Name Evolution:

Aeronautical Division, U.S. Signal Corps,	1907 – 1914
Aviation Section, U.S. Signal Corps,	1914 – 20 May 1918
Air Service, U.S. Army,	1918 – 02 July 1926
Army Air Corps,	1926 – 20 June 1941
AAC then became the combat arm of the AAF till 1947	
Army Air Force,	1941 –18 September 1947
U. S. Air Force	1947 – today.

Till they met Fletcher

JAPANESE ATTACK FORCE - VAdm Naguma
CARRIERS
    Akagi, *flag*   Kaga   Soryu
    Hiryu   Zuikaku   Shokaku

SUPPORT FORCE - VAdm Mikawa
Battleships
    Hiei   Kirisima

Heavy Cruisers
    Tone   Chikuma

SCOUTING FORCE - RAdm Omori
    Abukuma, light cruiser, flagship of the destroyer flotilla

Destroyers
    Tanikaze   Hamakaze   Urakaze
    Kasumi   Arare   Kagero
    Shiranuhi   Akigumo   Isokaze

Midway Raid
    Ushio   Akebono

SUPPLY FORCE.
Tankers
    Kyokuto, *flag*
    Kenyo   Kokuyo   Shinkoku
    Toho   Toei   Nippon

SUBMARINE FORCE
Three fleet reconnaissance : *I-19, I-21, I-23* travel with the strike fleet and are the only ones the public seems to know about and considers part of the attack. There were many more.
Five attack submarines with five midget submarines strapped aboard:
    I-16 with A , I-18 w/E , I-20 w/D , I-22 w/B , I-24 w/C.
The five midget subs, each capable of penetrating Pearl Harbor, at least one did, and each was capable of firing two torpedoes.
Twenty other subs were moved into Hawaiian waters.
    Squadron 1: *I-9, I-15, I-17, I-25.*
    Squadron 2: *I-1, I-2, I-3, I-4, I-5, I-6, I-7.*
    Squadron 3: *I-8, I-68, I-69, I-70, I-71, I-72, I-73, I-74, I-75.*
Two others were sent directly to the U.S. West Coast:
    I-10 and *I-26.*
The other half of the submarine fleet was sent to support West Pacific attacks.

Japanese Aircraft

Distribution of the attacking planes:

Carrier	A6M "Zero"	B5N "Kate"	D3A "Val"	Total
Akagi	18	27	18	63
Kaga	18	27	27	72
Soryu	18	18	18	54
Hiryu	18	18	18	54
Shokaku	18	27	27	72
Zuikaku	18	27	27	72
Total	108	144	135	387
w/spare	(126	162	153	441)

"Zero" – fighter
"Kate" – torpedo/horizon bomber
"Val" – dive bomber

Three aircraft of each type on each carrier were carried disassembled as spares, a total of 54 spare aircraft.

Combat air patrols were flown over the fleet by 48 Zeros.

Six scouting seaplanes from the cruisers flew reconnaissance over Pearl Harbor and the sea surrounding the attacking fleet. *Tone* and *Chikuma* were specifically designed as seaplane carrying, heavy cruisers.

Admiral Fletcher in Continuous Combat
Dec 7, 1941, to Sept 30, 1942
Commander of Heavy Cruiser Force 6 at Pearl Harbor.

Half of Cruiser Six was in dry-dock getting upgrades in anticipation of the coming fray. The other two were at sea.

Minneapolis (CA-36) departed Pearl Harbor Dec 5, 1941, for operational and gunnery practice about 20 miles south of Oahu with a group of recently arrived WWI destroyers converted to high-speed minesweepers. On word of the attack December 7th, she joins the search for the Japanese fleet. *Minneapolis* is sighted and reported as an enemy aircraft carrier ; bombers are sent which recognize the U.S. ship. She returns to Pearl Harbor on Dec 10.

Astoria (CA-34) sortied December 5th with RAdm Newton's TF 12 built around *Lexington* (CV-2) to deliver Marine fighter aircraft to Midway Island. It is too late. The ferry mission is canceled with orders to join up with *Indianapolis* (CA-35, VAdm Brown) which had been on a simulated bombardment of Johnson Island with another group of fresh destroyer minesweepers. They search southwest of Oahu "to intercept and destroy any enemy ship in the vicinity of Pearl Harbor. . . ." They reenter Pearl Harbor on 13 December.

Fletcher returns to the confusion of Pearl Harbor and prepares to put the heavy cruiser division Six to sea on the 16th to rendezvous with and screen an emergency convoy to Wake Island.

Attempted Relief of Wake Island

The anticipated Japanese invasion of Wake Island on December 11th is repulsed by Marines on the island. Shore battery gunfire sinks destroyer *Hayate* and damages four others. USMC F4F Wildcats bomb and sink destroyer *Kisaragi* and strafe and damage two light cruisers. Following the abortive assault, the Japanese bomb daily with flying boats in pre-dawn raids and, later in each day, land attack planes bomb Wake.

The American public is encouraged by this victory.

Admiral Kimmel sets up to rescue Wake Island. On the 14th, he orders VAdm Brown to take *Lexington* and escorts as Task Force 11 to sail for the Marshall Islands to disrupt the suspected source of the attacks on Wake.

Next day, a relief convoy around seaplane tender *Tangier* (AV-8) departs Pearl as *Saratoga* (CV-3) arrives from the West Coast to refuel.

On the 16th Fletcher's cruiser division escorts *Saratoga* as Task Force 14 to catch up with the convoy that is to return 1,145 civilian contractors. On that same day, Japanese Pearl Harbor Attack Force dispatches carriers *Hiryu* and *Soryu*, with two heavy cruisers and two destroyers as escorts to reinforce a planned second attack on Wake Island. Japanese naval land attack planes continue to bomb Wake. VAdm Halsey returns with *Enterprise* from searching for the Japanese attackers

Admiral Kimmel is relieved on the 17th by Secretary of the Navy Knox, replaced temporarily with VAdm Pye, commander of the no longer existent Battle Force, pending the arrival of the new Pacific Fleet Commander, newly frocked Admiral Chester Nimitz.

Next day Pye orders *Lexington* (CV-2) north to join with *Saratoga*.

On the 20th, Halsey's TF 8 with *Enterprise* proceeds to waters west of Johnston Island to cover TF 11 and TF 14 operations and Hawaii.

On the 21st and 22nd, planes from carriers *Soryu* and *Hiryu* and land planes bomb Wake. The last two flyable USMC F4Fs intercept the raid. One is shot down, the other is badly damaged. Fletcher's *Saratoga* TF 14 is ordered to wait for *Lexington* TF 11. Fletcher refuels for whatever is to come.

Dec 23. Two Japanese high speed transports intentionally run ashore to facilitate landing of naval troops. Planes from the two carriers as well as seaplane carrier *Kiyoawa* provide close air support for the invasion. Wake Island is captured after a gallant resistance offered by the garrison that consists of marines, sailors, volunteer civilians and a USAAF radio detachment.

Uncertainty over the positions of and the number of Japanese carriers and reports that indicate Japanese troops have landed on the atoll compel VAdm Pye to recall the relief mission.

During the return, *Saratoga* flies off USMC F2A Buffalo fighters to Midway. *Tangier* disembarks defensive troops and a ground echelon for the fighters.

January 1, 1942. Fletcher takes command of Task Force 17 centered on USS *Yorktown* (CV-5) returning to the Pacific Fleet after eight months of service on Neutrality Patrols in the Atlantic. His first task is to escort troopships with Marines to the South Pacific. This gives him a grounding in aircraft carrier operations. He raids Japanese held islands on the way back to Pearl Harbor.

Till they met Fletcher

Battles in the Java Sea.
East Indies are lost

During this period, on the other side of the Pacific, the Japanese are rapidly moving down the East Indies from their earlier conquests in Indochina and the Philippines. Borneo - Jan 10, Bismarck Archipelago (Rabaul) - Jan 23, Celebes - Jan 24, New Guinea - Feb 9, Sumatra - Feb 16, Bali and air raid on Darwin - 19 Feb, Timor - Feb 20.

A joint command of American, British, Dutch, and Australian, ABDA, General Sir Wavell, with a task force of cruisers and destroyers under Dutch Rear Admiral Karel Doorman attempts to attack invading troopships. These had some success at Macassar Strait on Jan 24, but the fleet is lost in battles around Java : Badoeng (Lombok) Straits, Feb 18-19 ; Java Sea, Feb 27 ; and Sundra Strait, Feb 28. ABDA command is dissolved March 1 and the Americans assume responsibility for the whole of the Pacific.

In April, Burma and the Philippines are captured and the Japanese Navy raids Ceylon and the Indian Ocean east coast of India sinking the small carrier, **Hermes**, heavy cruisers **Dorsetshire** and **Cornwall,** and 23 merchant ships pushing the British out of the Pacific and Indian Oceans back to Africa.

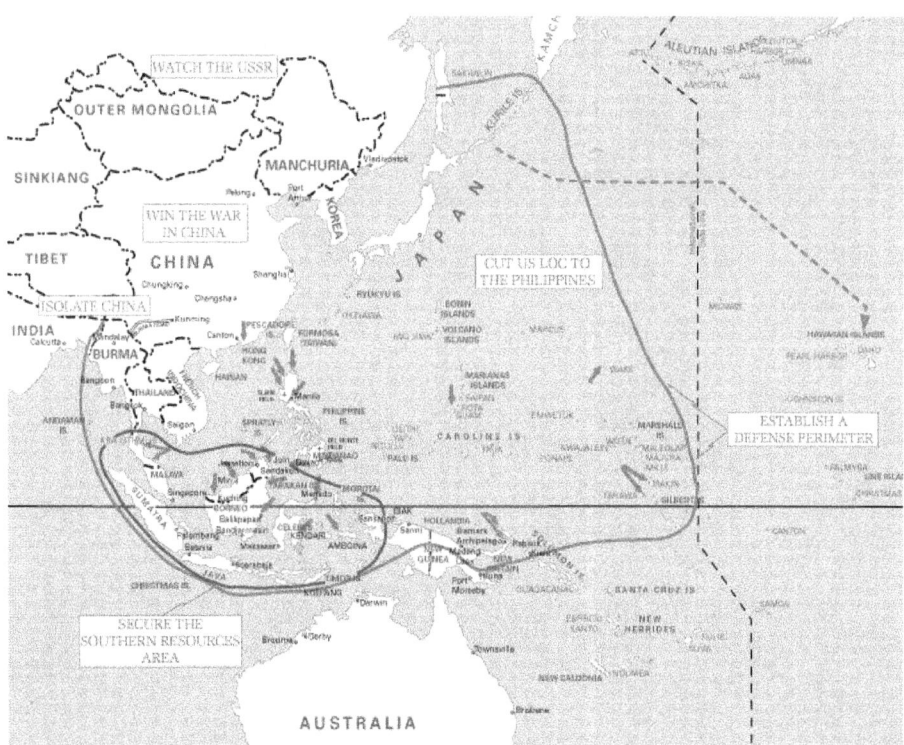

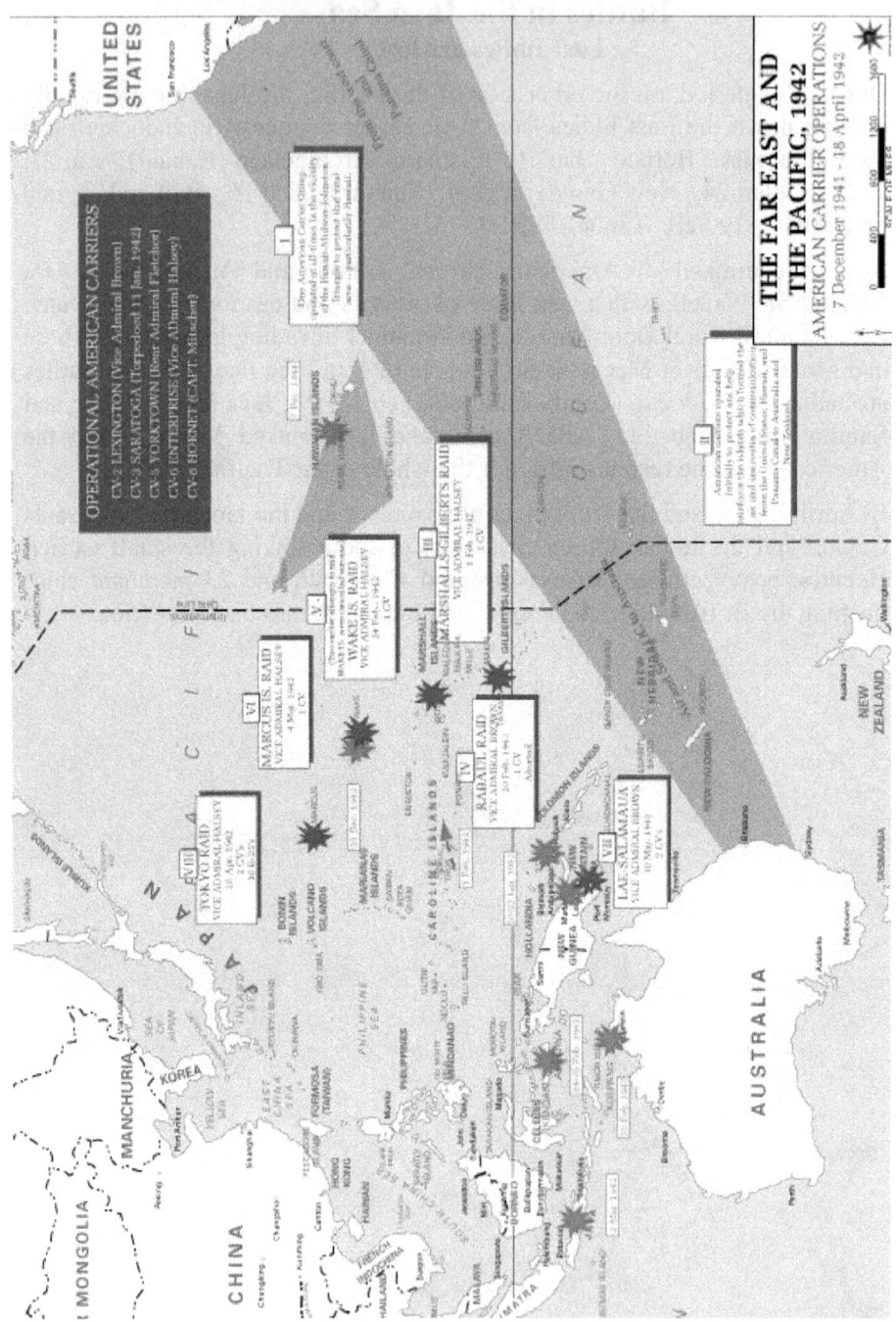

Carrier Movements in the First Year
1941
Dec 8. The carriers are at sea and not attacked at Pearl Harbor where 8 battleships are sunk or damaged. By the end of the first year, four of the seven carriers the U.S. had for both oceans are sunk in the Pacific.

- *Lexington* (CV-2) is on her way with TF 12 to deliver planes to Midway. Turns away to search for a Japanese fleet.
- *Saratoga* (CV-3) is on the West Coast having just completed overhaul. Immediately loads with planes for transport to Wake Island.
- *Enterprise* (CV-6) is returning from delivering aircraft to Wake Island.

There are four carriers in the Atlantic on "Neutrality Patrol."

- *Ranger* (CV-4) is returning to Norfolk from an ocean patrol near Trinidad.
- *Yorktown* (CV-5) is at Norfolk after another Neutrality Patrol.
- *Wasp* (CV-7) lays at anchor in Grassy Bay, Bermuda, part of Neutrality Patrol.
- *Hornet* (CV-8) commissioned six weeks ago and is training out of Norfolk.
- Plus, *Long Island* (AVG-1), the new auxiliary carrier operates out of Norfolk, conducting experiments to prove the feasibility of aircraft ferry operations from converted cargo ships.

Dec 12. *Yorktown* scheduled for maintenance, stops just long enough to get most .50" mg replaced with 20mm and a tweak to her radar, loads aircraft and supplies.
Dec 13. *Lexington* returns to Pearl Harbor.
Dec 14. *Lexington* (VAdm Brown, TF 11) sails for Jaluit, Marshall Islands to relieve pressure on Wake Island.
Dec 16. *Yorktown* sails for Pacific.
Dec 16. *Saratoga* (RAdm Fletcher) sails to relieve Wake Island.
Dec 19. *Enterprise* (VAdm Halsey) sorties to cover *Lexington* and *Saratoga* actions.
Dec 20. *Lexington* raid canceled, ordered to join *Saratoga* to relieve Wake Island.
Dec 23. Wake Island captured; carriers recalled.
Dec 25. *Saratoga* flies off aircraft to Midway.
Dec 27. *Lexington* and *Saratoga* return to Pearl Harbor.
Dec 30. *Yorktown* arrives San Diego.

1942

Jan 1. TF 17 (RAdm Fletcher) formed on *Yorktown* (CV-5).
January. *Lexington* (Brown) patrols Oahu-Johnston-Palmyra triangle.
Jan 6. *Yorktown* (Fletcher) sails from San Diego to escort Second Marine Brigade to American Samoa.
Jan 11. *Enterprise* (CV-6), TF 8 (VAdm Halsey) departs Pearl to join escort of Marines to Samoa.
Jan 12. *Saratoga* (VAdm Leary) TF 14 torpedoed 500 miles SE Hawaii ; able to return to Pearl and is sent to West Coast for repairs until June 1.
Jan 20. Marines arrive Pago Pago covered by *Yorktown* and *Enterprise*.
Jan 21. The two task forces set course for the Japanese-held Marshalls and Gilberts.
Jan 22. *Lexington* (CV-2) TF 11 (VAdm Brown) departs to raid Wake Island.
Jan 23. *Neches* (AO-5), fleet oiler, torpedoed and sunk. *Lexington* has to cancel raid on Wake Island.
Jan 30. *Lexington* (Brown) sent south to cover the attack on the Gilberts and Marshalls.
Feb 1. *Enterprise* (Halsey, TF 8) bombs Kwajalein and Spruance's cruisers hit Wotje in Marshalls. *Enterprise* damaged.
Feb 2. *Yorktown* TF17 (Fletcher) strikes Makin, Mili and Jaluit in Gilberts.
Feb 5. *Enterprise* returns Pearl Harbor.
Feb 6. *Yorktown* returns Pearl Harbor.
Feb 14. *Enterprise* (Halsey) leaves Pearl Harbor towards Wake Island.
Feb 15. *Yorktown* (Fletcher) leaves Pearl Harbor to patrol South Pacific.
Feb 16. *Lexington* TF 11 (VAdm Brown) heads for an attack on Rabaul, New Britain.
Feb 19. *Lexington* (Brown) is discovered and fights off an air attack and cancels air raid on Rabaul.
Feb 21. *Lexington* patrols Coral Sea while trying to refuel.
Feb 24. *Yorktown* to New Hebrides to meet *Lexington* for another Rabaul raid.
Feb 24. *Enterprise* (Halsey) raid on Wake Island.
Mar 4. *Enterprise* (Halsey) raids Marcus Island.
Mar 6. *Lexington* (Brown) and *Yorktown* (Fletcher) join up intending raid on Rabaul, but are diverted when, on
Mar 8. Japanese invade New Guinea at Lae and Salamaua.
Mar 10. *Lexington* and *Yorktown* attack the Japanese invasion force at New Guinea, sinking four ships and damaging 13 out of 18.
Mar 11. *Enterprise* (Halsey) returns to Pearl Harbor.
Mar 12. Fall of Netherlands East Indies.
Mar 15. *Yorktown* in Coral Sea ; *Saratoga* on W. Coast for repairs ; *Lexington* to Pearl ; *Enterprise* for overhaul. *Hornet* en transit from Norfolk.

Till they met Fletcher

Mar 20. *Hornet* (CV-8) arrives San Francisco.
Mar 26. *Lexington* arrives Pearl to drydock for overhaul.
Apr 4. *Hornet* leaves San Francisco with Doolittle's B-25s.
Apr 8. *Enterprise* (Halsey) departs to rendezvous with *Hornet*.
Apr 14. Nimitz assigned to South Pacific. Sends *Yorktown* (Fletcher) to Tongatabu, south of Samoa to replenish for anticipated defense of New Guinea and Solomon Islands at the end of the month.
Apr 15. *Lexington* TF 11 (RAdm Fitch) leaves Pearl to join *Yorktown* (Fletcher) for raid on Rabaul.
Apr 18. Doolittle Raid on Tokyo from *Hornet* escorted by *Enterprise*.
Apr 20. *Lexington* (Fitch) arrives South Pacific.
Apr 22. *Lexington* joins *Yorktown* at Tongutabu.
Apr 25. *Enterprise* and *Hornet* return to Pearl Harbor from Tokyo raid.
Apr 27. *Yorktown* heads for Coral Sea.
Apr 29. Japanese occupy former RAAF base at Tulagi, Solomon Islands.
Apr 30. *Enterprise* and *Hornet* speed towards Coral Sea ; too far to reach in time. Patrol South Pacific till mid-May.
May 1. *Yorktown* (Fletcher) and *Lexington* (Fitch) join up.
May 4. While *Lexington* (Fitch) refuels, Fletcher takes *Yorktown* to attack Japanese invasion force at Tulagi sinking several small ships and damaging several others.
May 7. **Battle of Coral Sea.** Fletcher sends his Australian cruisers to block Port Moresby, PNG, invasion fleet at Jomard Passage, Louisiade Archipelago. *Lexington* attacks and sinks the support carrier causing the invasion force to withdraw. Meanwhile, a covering fleet with two fleet carriers closes in from his rear and destroys his oiler and its destroyer. *Yorktown* fighters fend off an evening attack.
May 8. Two fleets of two carriers each simultaneously attack each other the next morning. Japanese carrier *Shokaku* is damaged while *Zuikaku* escapes in a squall but with her air group severely damaged. Both U.S. carriers are damaged. A later explosion cripples **Lexington** which is evacuated and sunk. *Yorktown* needs to return to Pearl for repair and participation in Midway.
May-mid. *Enterprise* and *Hornet* patrol South Pacific.
May 22. *Saratoga* departs Puget Sound for San Diego.
May 26. *Enterprise* and *Hornet* return to Pearl from South Pacific.
May 27. *Yorktown* arrives Pearl for repair.
May 28. TF 16 *Enterprise* and *Hornet* (RAdm Spruance) sortie to Midway.
May 30. TF 17 *Yorktown* (RAdm Fletcher) to Midway after rushed repairs.

June 1. *Saratoga* heads for Pearl Harbor.
June 4. **Battle of Midway** – the decisive battle of the Pacific. **Four enemy carriers** sunk by three U.S. carriers.
June 5. *Long Island* (AVG-1) arrives San Francisco, joints TF 1 of battleships on West Coast as scout.
June 6. Battle damaged *Yorktown* sunk by submarine *I-168*.
June 7. Too late for Midway, *Saratoga* replenishes *Enterprise* and *Hornet* aircraft which head briefly for Alaska.
June 10. *Wasp* (CV-7), *North Carolina* (BB-55), *Quincy* (CA-39), *San Juan* (CLaa-54) and 6 DD arrive Pacific to became TF 18 (RAdm Noyes).
June 13. *Saratoga, Enterprise, Hornet,* return to Pearl Harbor.
June 19. Adm Robert L. Ghormley assumes command of South Pacific.
June 22. *Saratoga* (RAdm Fitch) ferrys planes to Midway, returns June 29.
June 26. Fletcher promoted to Vice Admiral leading *Saratoga* [flag], *Wasp*, and *Enterprise* to South Pacific.
July 1. *Wasp* (TF 18, Noyes) departs San Diego for Tonga Islands escorting five troopships of Marines.
July 7. *Saratoga* (TF 11, VAdm Fletcher) leaves for South Pacific.
July 9. *Hornet* (CV-8) has new radar installed and trains out of Pearl Harbor.
July 15. *Enterprise* (TF 16, RAdm Kinkaid) leaves for South Pacific.
July 17. *Long Island* returns to West Coast to resume training pilots.
July 18. *Wasp* arrives Tongatabu.
July 28. Fleet rehearsal for assault on the Solomon Islands held at Fiji.
Aug 2. *Long Island* (ACV-1) departs Pearl to ferry planes to Solomons.
Aug 7. Marines land at Tulagi (sharp fight) and Guadalcanal (no resistance).
Aug 9. Occupation complete. Carriers withdraw.
Aug 9. Savo Island – U.S. screening force wiped out. Navy transports retreat.
Aug 17. *Hornet* (TF 17, RAdm Murray) departs Pearl for Solomons.
Aug 20. *Long Island* flies off planes to Henderson Field, Guadalcanal.
Aug 23. Japanese start to reinforce Guadalcanal; fight until Feb 7, 1943.
Long Island arrives Efate, New Hebrides.
Aug 24. **Battle of Eastern Solomons**. Major Japanese strike fleet turned back with massive loss of aircraft on both sides. *Saratoga* sinks ***Ryujo***. *Enterprise* is damaged, returns to Pearl Harbor.
Aug 31. *Saratoga* hit by a torpedo from *I-26*; out of action for the next three months.
Sept 5. *Copahee* (ACV-12) departs Alameda, CA, for Noumea, New Caledon.
Sept 7. *Hornet* avoids two torpedoes destroyed by depth charge from TBF.
Sep 10 to Oct 16. *Enterprise* repairs at Pearl Harbor.
Sep 12. *Wasp* delivers aircraft to Henderson Field, Guadalcanal.
Sep 15. ***Wasp*** hit with three torpedoes by *I-19* and sunk while screening troops from New Hebrides to Guadalcanal.

Till they met Fletcher

Sep 16. *Hornet* is only active carrier left in Pacific for over a month.
Sep 21. *Saratoga* arrives Pearl for repair for 6 weeks. Fletcher returns to Washington, DC.
Sep 28. *Copahee* (ACV-12) arrives at Noumea with a cargo of planes, stores and passengers.
Oct 5. *Hornet* (TF 17, Murray), hit targets in the Buin-Tonolei-Faisi area of Papau New Guinea.
Oct 10. *Nassau* (ACV-16) arrives at the Naval Air Station, Alameda, CA, to load aircraft.
Oct 11. *Copahee* launches 20 Marine fighter planes for Henderson Field.
Oct 13. Army troops start to relieve Marines on Guadalcanal.
Oct 14. *Nassau* departs Naval Air Station, Alameda, California for Pearl Harbor with aircraft for the South Pacific.
Oct 16. *Enterprise* (TF 16, Kinkaid), repaired at Pearl, departs once more for the South Pacific.
Oct 16. *Hornet* attacks beached Japanese transports on Guadalcanal; destroys seaplanes at Rekata Bay, Santa Isabel Island.
Oct 18. Halsey takes over South Pacific forces from Ghormley.
Oct 24. *Hornet* and *Enterprise* join up in New Hebrides.
Oct 26. **Battle of the Santa Cruz Island.** *Enterprise* and *Hornet* repulse a force twice their size. ***Hornet*** is lost ; *Enterprise* damaged.
Oct 29. *Copahee* arrives San Diego for an overhaul.
Oct 30. *Enterprise* enters Noumea, New Caledonia, for repairs. USN briefly has *no* active fleet carriers in the Pacific. *Nassau* (AVG-16) arrives Palmyra Island.
Nov 3. *Altamaha* (ACV-18) departs San Diego for South Pacific.
Nov 10. *Saratoga,* repaired, departs Pearl for Fiji.
Nov 11. *Enterprise* sails for Solomons, repair crews still aboard.
Nov 13. **Naval Battle of Guadalcanal**. *Enterprise* planes disable battleship *Hiei*.
Nov 14. *Enterprise* planes help sink heavy cruiser *Kinugasa*.
Nov 15. *Enterprise* returned to Noumea to complete her repairs
Nov 24. *Altamaha* delivered part of her cargo at Espiritu Santo, New Hebrides, and continued on to New Caledonia.
Nov 30. **Battle of Tassafaronga**, U.S. cruiser force destroyed by Tokyo Express.
Dec 4. *Enterprise* (RAdm Sherman) trains out of Espiritu Santo, New Hebrides.
Dec 5. *Saratoga* (RAdm Ramsey) arrives Noumea.
Dec 10. *Saratoga* patrols eastern Solomons for the next year.

Year-end: Dec 31, 1942.

In the Pacific
Four of the five great carrier battles of WWII have been fought – Coral Sea. Miway, Eastern Solomons, and Santa Cruz..

Fleet carriers: *Lexington* (CV-2), *Yorktown* (CV-5), *Wasp* (CV-7), and *Hornet* (CV-8) have been sunk.
Saratoga (CV-3) and *Enterprise* (CV-6) patrol between New Hebrides and the Solomons escorting supplies to Guadalcanal.

In the Atlantic.
Ranger (CV-4) was in overhaul at Norfolk after North African operations.
Santee (ACV-29) after N. Africa, was in Trinidad on way to South Atlantic.
Charger (AVG-30) area of operations throughout the war was Chesapeake Bay.
Essex (CV-9) was commissioned this day at Newport News, destined for the Pacific in mid-1943.

Auxiliary Carriers:
Long Island (ACV-1) trained carrier pilots at San Diego.
Copahee (ACV-12) was in overhaul at San Diego after a delivery to S. Pacific and flying off 20 Marine fighters to Guadalcanal.
Nassau (ACV-16) operated on the convoy route between Palmyra, New Caledonia and New Hebrides.
Altamaha (ACV-18) was at Espiritu Santo for training after delivery of planes.
Sangamon (ACV-26), *Suwannee* (ACV-27), *Chenango* (ACV-28), after participation in the invasion of North Africa, were en-route to the Pacific.
Bogue (ACV-9) was on an extensive shakedown and repair period from Tacoma, for Atlantic.
Card (ACV-11) had just been commissioned at Tacoma and destined for the Atlantic.
Core (ACV-13) had just been commissioned at Tacoma and destined for the Atlantic.
Bogue, Card, and *Core* will each have a successful anti-submarine career in the Atlantic with their task groups sinking 32 submarines.

Auxiliary aircraft carriers, ACV, will be re-designated as combat ships, CVE, escort aircraft carriers in mid-1943. One hundred twelve were built in U.S. with eleven more building at war end. Thirty-nine went to the U.K. as lend lease, until they ran out of sailors to crew them.

The Battle of the Coral Sea -- 4-8 May 1942

Background
The Japanese purpose of the War in the Pacific was to obtain resources to continue her conquest of China. To this end she attacked the major power centers -- the U.S. at Pearl Harbor, took the Guam, Wake and other Central Pacific outposts ; invaded S.E. Asia to cut off resupply from India to China ; took the resources of the Dutch East Indies while destroying the ABDA naval forces ; took the Philippines and Malaya ; and knocked the British fleet from India back to Africa.

The next logical steps are to remove Australia from the war, and then to isolate Hawaii to limit America to the continental U.S.

However, the symbolic bombing of Tokyo by the Doolittle Raid of B-25s launched from a carrier, even though doing little damage, changed the Japanese priorities in the Pacific.

Original Plans
On return of the Japanese fleet from the Indian Ocean, an attack to the south was planned to:
- Complete the conquest of New Guinea, which opens Australia
- to air attack,
- Complete taking the Solomon Islands which thrust into the shipping lane from America
- Raid Australia with the full carrier fleet to destroy all offensive resistance forces, although the Japanese Army was unprepared to occupy Australia.

At this point all of the original goals have been met with more ease than expected, so the plan was expanded to occupy another ring of islands to extend the defensive parameter.

New Plans
- The attack south was changed from a six, fleet carrier strike with supporting battle and occupation forces, to a two fleet carrier support for the movement of troops around to the south side of New Guinea and to complete occupation of the Solomons with air bases that would be able to interdict convoys to Australia. The other carriers had to prepare for an attack on Midway.

- The isolation of Hawaii was moved up and became a grand scheme to destroy the remaining U.S. forces in the Pacific by an attack on Midway, the western-most of Hawaiian archipelago, and on Alaska, expecting to defeat the U.S. forces at sea. Nine carriers, one training carrier, the battle fleet, and lesser types, in total 124 warships were assigned to the task to invade Midway, to damage U.S. carriers, and to let the battle fleet complete the destruction. As it turned out, two of Japan's six fleet carriers were to miss the fun because of damages received at the Coral Sea. c
- After the expected Japanese victory at Midway, forces would deploy to about 175°W latitude and accept peace while retaining their conquests and continue with the takeover of China.

The Setup - U.S.
After the disaster at Pearl Harbor and Wake Island and the acceptance of inability to reinforce the Asiatic fleet and its assured destruction, the entire theater was in organizational shuffling. With the new year, Adm Nimitz took over the Pacific from Admirals Kimmel / Pye.
- *Lexington* (CV-2, VAdm Brown, TF 12) stood guard for Hawaii. *Saratoga* (CV-3, VAdm Leary, TF 11) was soon torpedoed in the same area and put out of action and missed both the Battles of Coral Sea and Midway. RAdm Fletcher took newly arrived *Yorktown* (CV-5, TF 17) to escort Marines to Samoa, joined by *Enterprise* (CV-6, VAdm Halsey, TF 8). These two task forces raided the Marshalls and Gilberts on their way back to Pearl.
- The two Vice Admirals headed out for raids. *Enterprise* hits Wake and Marcus. *Lexington* is attacked before she can strike Rabaul and stays to patrol in the Coral Sea. *Yorktown* goes to the South Pacific to protect convoys now arriving from the West Coast. *Lexington* and *Yorktown* join up for another attempt at Rabaul, but they divert to attack the Japanese landing on the north coast of New Guinea. *Lexington* heads back to Pearl for overhaul. *Enterprise* and newly arrived *Hornet* (CV-8) make the Doolittle Raid on Japan.
- RAdm Fitch replaces VAdm Brown in *Lexington*, with Brown moved to Amphibious Force Pacific Fleet. Adm. Nimitz personally takes over South Pacific for a short time -- to show equal rank with Gen. MacArthur in setting up jurisdictions between the Army in SW Pacific and the Navy in South Pacific.

Till they met Fletcher 93

- *Lexington* rejoins *Yorktown* at Tonga (SW of Samoa) intending to make another try on Rabaul.
- Reports come in of Japanese carriers arriving at Rabaul. *Yorktown* heads for the Coral Sea followed by *Lexington*. The two task forces have five heavy cruisers, nine destroyers, supplied by two tankers each with a destroyer. VAdm Ghormley comes over from Europe to take South Pacific, Nimitz returns to Pearl and sends *Enterprise* and *Hornet* towards the Coral Sea (they don't arrive in time for the Battle). MacArthur's Navy, under RAdm Crace, RN, with three cruisers : HMAS *Australia* (CA.84), heavy; HMAS *Hobart* (CL.63), light ; USS *Chicago* (CA-29) ; and two U.S. destroyers also came to the defense of Australia.

The Setup – Japanese
The Japanese attack in the Coral Sea has three task forces.
- Port Moresby Task Force to swing around the east coast of Papua New Guinea and to take the capital on the south coast, opposite Australia.
- An occupation force to reach the southern-most Solomons, to immediately establish a seaplane base at Tulagi, and then to build an airfield on Guadalcanal across the sound.
- A carrier strike force of two fleet carriers, two heavy cruisers, and two destroyers to wipe out any Allied naval forces that tried to interfere. Then to strike bases in northeastern Australia in support of the Port Moresby invasion.

The Opening Sequence
- On 4 May *Yorktown* is on station in the Coral Sea, Fletcher responds to the Japanese occupation force at Tulagi, while *Lexington* task group continues refueling from her 1,700 mile trip up from Tonga. Minor damage was committed at Tulagi, insufficient to interfere with setting up a seaplane base, but the U.S. thought it had sunk most of the Japanese ships (which had simply sailed away between air raids).
- *Lexington* and *Yorktown* rejoin, refuel the combined task force, and send the tankers out of the way. Both sides search for the other. They miss, though at one point are only 70 miles from each other in bad weather.

On 7 May each side detects elements of the other.
- Japanese search planes find U.S. oiler, *Neosho*, and destroyer, *Sims*, steaming to their next rendezvous point. Mistaking them for a carrier and cruiser, they launch a full and fierce attack, sinking the ***Sims*** and wrecking ***Neosho***.
- Meanwhile, U.S. search planes find the Port Moresby invasion force. Thinking it is the strike fleet, they launch a full attack. Small escorting carrier ***Shoho*** is destroyed. "Scratch one flattop."
- The Port Moresby invasion force halts, waiting for the Allied forces to be destroyed before venturing further.
- Fletcher sends Crace's cruiser force to block the Japanese troop convoy, while *Yorktown* and *Lexington* continue to seek the Japanese carrier strike force.
- Aircrews encounter each other with the loss of nine Japanese bombers.

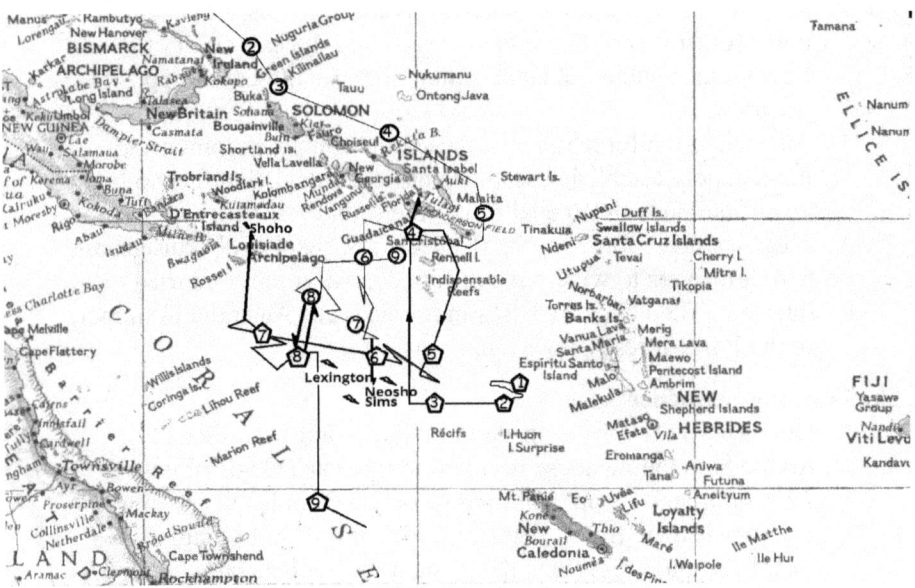

Circle = IJN. Pentagon = USN. Number is date. Location at noon.

On dawn 8 May, each side finds the other and launch simultaneous attacks.
- U.S. planes find *Shokaku* and make three bomb hits putting her deck out of action. *Zuikaku* was not found as she hid in a rain squall. She landed the remaining Japanese planes from both Japanese carriers *Shokaku* retired to Rabaul before going on to Japan for repairs which take three months. *Zuikaku* returns to Japan to replenish her aircrews, which takes a month. Both big carriers miss the Battle of Midway.
- The Japanese invasion force, now without air cover, returns to Rabaul.

- The two American carriers were each hit : *Yorktown* with one bomb, *Lexington* by two torpedoes and two bombs. Later, **Lexington** suffered a gasoline explosion and was abandoned. *Yorktown* retired and returned to Tonga and then to Pearl for repair.
- Crace's Cruisers fought off Japanese land-based aircraft in their forward role of blocking the Japanese invasion force. They were even attacked by B-17s from Australia, but they held on until well after the Japanese withdrew.
- *Enterprise* and *Hornet* arrived after the battle and patrolled east of the Coral Sea, scaring off a Japanese occupation force sailing for the Phosphate Islands, until recalled for the Battle of Midway less than a month after the Battle of the Coral Sea.

Conclusion
- This battle was the first of the carrier battles of the Pacific War. Ships from neither side sighted the other. The Japanese had fought at Pearl Harbor, Java, and Indian Ocean and should have been the better prepared. Weather made finding the Japanese difficult, while the Americans were in sunshine. Radar played little part. Both sides had an opportunity to learn about damage control on carriers, but not soon enough for Midway the following month. Tactically, the two forces were almost evenly matched. Yet the Japanese lost more planes.
- The Japanese lost a small carrier, damage to a fleet carrier, and significant loss of aircrews and planes. This was a tactical victory over the Americans who lost a fleet carrier, a tanker, and a destroyer and damage to another fleet carrier. Tonnage loss favored the Japanese.
- Strategically, however, the Japanese objective was thwarted ; the invasion fleet withdrew and never returned to threaten Australia. New Guinea becomes an infantry battle that lasted for two and a half years.
- The big picture will progress next to the Battle of Midway/Aleutians in one month where the nature of the war changed.
- The Japanese now had four active fleet carriers plus four lighter carriers, to the U.S. three fleet carriers.
- Japan had two fleet carriers out of action in Japan and the U.S. had the fleet carrier, *Saratoga,* repairing at Puget Sound. *Yorktown* would briefly repair at Pearl in time to make it to Midway. The Battle of the Coral Sea had taken strategic victory from a tactical defeat, but the Allies were still outnumbered and still losing the Pacific War.

USS LEXINGTON (CV-2)

IJN SHOHO "Scratch one Flattop"

Cruce's Flagship, HMAS **AUSTRALIA**   heavy cruiser

Photo # 80-G-640553 USS Yorktown operating at sea, April 1942

ZUIKAKU

SHOKAKU
Each 29,800 tons; 845 feet; 34.5 knots; 72 aircraft

The Battle of the Coral Sea – May 1942
Naval Forces Engaged

Type	Japanese	Allied
Fleet Carriers CV	2	2
Light Carriers CVL	1	0
Heavy Cruisers CA	6	7
Light Cruisers CL	3	1
Destroyers DD	15	13
Gun Boats PG	6	0
Oilers AO	3	2
Transports AK	13	0
Auxiliary	13	1
Aircraft	139	136

Japanese Forces

"MO" OPERATION
  VAdm Inoue Shigeyoshi
  (in *Kashima*, at Rabaul)
"MO" STRIKE FORCE
  VAdm Takagi, CarDiv5
Zuikaku   CV   64 a/c
  RAdm Hara Chuichi
Shokaku   CV   57 a/c
CruDiv5
Myoko   CA
Haguro   CA
Arike   DD
Yugure   DD
Shigure   DD
Shiratsuyu DD
Ushio   DD
Akebono   DD
Toho Maru AO Oiler
"MO" MAIN FORCE
  RAdm Goto Arimoto
CLOSE SUPPORT FORCE
  RAdm Goto
Shoho   CVL 18a/c

Allied Forces

Task Force 17
  RAdm Frank J. Fletcher
  (in *Yorktown*)
Task Group 17.5 (Carrier Group)
  RAdm Aubrey W. Fitch
  (in *Lexington*)
Lexington   CV-2 w/ 69 a/c
Yorktown   CV-5 w/ 66 a/c
Morris   DD-417
Anderson   DD-411
Hammann   DD-412
Russell   DD-414
Task Group 17.2 (Attack Group)
  RAdm Thomas C. Kincaid
Minneapolis   CA-36
New Orleans   CA-32
Astoria   CA-34
Chester   CA-27
Portland   CA-33
Phelps   DD-360
Dewey   DD-349
Farragut   DD-348
Aylwin   DD-355
Monaghan   DD-354

Till they met Fletcher 99

SUPPORT Force, Main Body
  RAdm Goto
Aoba (F)   CA
Kako   CA
Kinugasa   CA
Furutaka   CA
PORT MORESBY Invasion
ATTACK Force
  RAdm Sadamichi Kajioka
Yubari   CL
Oite   DD
Asanagi   DD
Mitsuki   DD
Mochizuki   DD
Yayoi   DD
Uziki   DD
1 Patrol Boat   PC
1 Auxiliary Craft
TRANSPORT Force
  RAdm Abe Koso
Tsugaru   CML
12 Transports   AK/AP
3 Minesweeepers
Goyo Maru   AO
Hoyo Maru   AO
Oshima   AR
CLOSE COVER Force
  RAdm Marumo Kuninori
Tenryu (F)   CL
Tatsuta   CL
Kamikawa Maru   AV
    seaplane tender
Nikkai Maru, Keijo Maru,
    Seikai Maru   PG
TULAGI INVASION Force
  RAdm Shima Kiyohide
Kikuzuki   DD
Yuzuki   DD
Okinoshima ML **
  Sunk by *S-42* May 11
Koei Maru   ML
Azumasan Maru   AK
Minesweepers (2); 3 MS; 2PC
Minelayers – 2
SUBMARINES
5 I-boats ; 2 RO-boats

Task Group 17.3 (Support Group)
  RAdm J.G. Crace, RN
Australia RAN   CA.84
Hobart RAN   CL.63
Chicago   CA-29
Perkins   DD-377
Walke   DD-416
Task Group 17.6 (Fleet Train)
Neosho   **AO-23**
Sims   **DD-409**
Tippecanoe   AO-21
Worden   DD-352
Task Group 17.9 (Search Group)
  (in Noumea)
Tangier   AV-8   w/12 PBY
SUBMARINES – 4
  S-38, S-42, S-44, S-47
** Sub *S-42* (SS-153) torpedoes
   minelayer, *Okinoshima,* May 11

BATTLE of MIDWAY – 4 June 1942

In brief.

Japan attacked Midway Island, near Hawaii, with their entire Navy. The U.S. heard they were coming and ambushed the Japanese. Four Imperial carriers were sunk and one American.

Conclusion.

The grand Japanese plan made by Yamamoto was good and Nugumo in command of the carriers did nothing wrong. Yet Japan had more misfortunes and Victory goes to the side that makes the fewer mistakes. The fighting forces of both sides showed exceptional bravery. American intelligence was better. Nimitz and Fletcher did everything well and Spruance did little wrong. Victory was to the U.S.N.

Japanese Plan.

The Japanese fleet was overwhelming, such that there were sufficient ships to make two attacks at once with each able to come to the aid of the other if needed. The main effort was at Midway, an island airbase within flying distance from Hawaii. The subordinate attack was on Attu and Kiska on the tip of the Aleutians Islands far out from Alaska pointing towards Japan. The goal of the engagement was to draw out the remaining U.S. carriers from Hawaii. Their assured destruction would allow the Japanese free reign in the Pacific. As long as the American carriers existed, the Japanese Navy had to retain a large guard force near the home islands to prevent Doolittle-type raids and required elements of the Imperial Navy to steam after every hit-and-run attack that Nimitz launched against outlying Japanese island posts. Once freed from such defensive roles, the entire IJN could go on the offense, capture New Caledonia, New Hebrides, Samoa in the South Pacific to isolate Australia, and take the scattered U.S. held islands in the Pacific Ocean, thus limiting the U.S. to the continental 48 States. There were even secret plans to later attack the Panama Canal to isolate the U.S. Pacific Coast from the Atlantic Fleet. And longer range hopes to cross the Indian Ocean to link-up with Rommel in North Africa.

The bait was an attack on Midway Island, the farthest extension of an island chain with Hawaii at the other end. An invasion fleet came up from Saipan escorted by a bombardment fleet of cruisers. A smaller invasion fleet went to the Aleutians escorted by a carrier task force to suppress American air forces. The principle attack was supported by the Japanese carrier strike force to put down all American resistance on Midway. A guard force of battleships was positioned between Midway and the Aleutians to go in either direction depending on where the Americans came out to fight. The main body followed the Japanese carriers, out of sight, with the largest battleship in the world has ever known, 18" guns, to finish off the American Navy whose largest ships available were heavy cruisers with 8" guns. There were some

Till they met Fletcher

old and repaired U.S. battleships operating out of San Diego which might have appeared, but they, too, would be overwhelmed by Japanese might. The plan was good.

American Plan

The American side had the advantage of having broken bits of the Japanese high-level naval codes. On indication that a major attack was being planned for the Central or North Pacific, the U.S. carriers, then all operating in the South Pacific, were recalled to Pearl Harbor. Once the details of the coming attack were determined, the American idea was to wait nearby, but out of sight northeast of Midway and attack the Japanese forces from the flank. The plan was of the highest risk.

Prequel

Leading up to this attack are two things of interest. The Doolittle Raid on mainland Japan two months before had pushed the Japanese into an early attempt to destroy the U.S. carriers. This meant that logistics were rushed. The Battle of the Coral Sea one month before had damaged one of the big Japanese carriers and decimated the air groups of it and another big Japanese carrier. Neither of these big carriers would be ready again in time for the attack on Midway. A submarine force of sixteen was assigned, most were sent to provide patrol lines to detect and weaken the American sortie from Hawaii. They could not leave in time, so arrived two days later than planned and the U.S. ambushing force had already passed through the area and was in place waiting for the Imperial Navy. Reconnaissance of Pearl Harbor by long-range flying boats was in the Japanese plan, a repeat of an earlier successful procedure of refueling from submarines at an isolated atoll in the Hawaiian chain. However, the U.S. had guessed that location and stationed ships there to prevent any repeat of that maneuver.

To Japanese advantage, their fleets moved through thick fog making detection difficult. Weather delayed refueling by a day but cleared just in time for the attack. The Alaskan force attacked first, the Midway force was a day later.

The Japanese Navy had two cruisers build for carrying scouting aircraft to accompany the carrier strike forces so that the carrier's fighting planes were not consumed in scouting and to make carrier operations free to concentrate all their efforts on attacking and defense. The American fleet was expected to be at Pearl Harbor and in the South Pacific so that the scouts did not expect to find anything in their 300 mile search pattern, a distance beyond American carrier attack range. The American carriers had to use combat aircraft for scouting duties, Fletcher did this in rotation so that only one U.S. carrier was delayed in the coming attack by having to wait for the recovery of scouting planes.

A key bit of Japanese bad luck. One of the cruiser launched scout planes flew over the American fleet in clouds and did not detect it. A second had trouble with its catapult and launched a half hour late and did not detect the U.S. fleet until after the American attack had been launched.

The Americans had a good idea from the code breakers of the time and direction from which the Japanese would attack. Midway Island had been reinforced with multi-engine scout planes (PBY) and bombers (B-17). Knowing where to look, the U.S. scouts found the Japanese and attempted to bomb them. But, bad for the Americans, a high flying bomber could not hit a maneuvering ship at sea. But it did divert the attention of the Japanese.

Japanese Action
The Japanese launched their air attack on Midway as planned. The island had been reinforced with fighter planes and antiaircraft guns. The outstanding Zero fighter decimated the American fighter defenses, but they were unable to suppress antiaircraft guns and the Japanese bombing results, though damaging, did not put the airstrip out of action. A succession of Navy, Marine, and Army Air Force planes from Midway attacked the Japanese carriers. Japanese Zeros and antiaircraft defenses prevented any damage to their ships.

American Action.
As the Japanese strike aircraft were returning to their ships, the American fleet was detected. Throughout the course of the morning, various Japanese scouting reports said there were from none or up to five U.S. carriers within range. As soon as one carrier was detected Nugumo turned to attack. However, he had two problems. All of his Zeros were engaged in defense of the carriers or had accompanied the strike force that was now returning. There were none to escort vulnerable torpedo planes on an attack. Second, he needed to land, refuel and rearm Zeros and while at it, he landed the returning

strike bombers, all of which he accomplished in only 15 minutes. The attack groups intended to strike the Americans were being lifted to the flight deck for launch when the first wave of American carrier torpedo bombers arrived. They were destroyed by the combat air patrol of Zeros. The American dive bombers arrived as the Zeros were finishing off the U.S. torpedo planes. Only ship's guns faced the dive bombers. Three squadrons from two U.S. carriers struck three Japanese carriers -- each was set afire from aviation gasoline and this set off bombs and torpedoes so that each carrier eventually sank.

Counter Attacks.

A fourth Japanese carrier was overlooked (two *Hornet* bomber squadrons got lost), but she was armed and her planes followed American bombers back to their carrier and, although detected by radar and half were destroyed, they managed to put a torpedo into *Yorktown* that caused damage, but she was still able to fight. They missed finding the other U.S. carriers. *Yorktown* had scouts out that found this fourth Japanese carrier and a second attack by U.S. forces succeeding in bombing her, but not before she had sent a second strike that hit *Yorktown* with two more torpedoes that caused flooding and fear of capsizing; she was abandoned.

End of the First Day.

Unable to fly aircraft at night, the American carriers retreated after hitting the fourth carrier. They steamed east till midnight before turning around to return to the battle scene by first light. The Japanese Navy was excellent at night fighting and the battleships that were coming to destroy the American fleet put on speed, but it soon became evident that they would never catch the Americans in the dark and would be subjected to air attack from both Midway and the U.S. carriers in the morning, so the Japanese withdrew to refuel and wait for the two carriers called from the Aleutians to arrive.

Second Day

One Japanese heavy cruiser from the invasion support fleet ran into another while evading a U.S. submarine. This saved the Japanese fleet. While the main fleet was stopped to refuel in the waters NW of Midway, the Americans from both Midway and the carriers concentrated on searching on incorrect sightings or search and rescue of downed pilots. The main Japanese battle force was not detected and a transport convoy loaded with troops and supplies each escaped. Two damaged enemy cruisers were found

Yamamoto tried two ways to salvage his plans. One was to lure the American carriers within range of Wake Island bombers and attack with his big battleships. Second, he sent some of his battleships and an arriving big carrier, *Zuikaku*, (refitted from the Coral Sea battle), north to Alaskan waters in case the American carriers went to attack those landing forces as hoped.

Third Day

Spruance scores on a cruiser.
One of the two damaged cruisers was finally sunk. The rest of the Japanese headed home to Hiroshima. The massive defeat was kept secret from the public and most of the armed services.

Salvage parties were working on *Yorktown* two days later when a submarine drawn to the battle sank her plus a destroyer alongside. ***Hammann*** (DD-412) sank at once, ***Yorktown*** the next day.

The Americans were smart enough not to go near the land-based planes on Japanese-held Wake Island and being unable to find the main Japanese fleet, withdrew to refuel and to be replenished with new aircraft from arriving carrier *Saratoga* and head for Alaskan waters. Nimitz soon realized that sending the carriers north to try to intercept the returning Alaska attack force would leave both Midway and Pearl Harbor vulnerable and he recalled the carriers within a few hours. The battle was over.

Results :
Four of Japan's six large carriers were sent to the bottom. One of four American carriers went down. The opposing fleets were now about evenly matched in air striking power. Japan's desire to not be tied down by chasing after the few American carriers and to free themselves for further offensive actions was now gone. The IJN was limited to defensive actions for the rest of the war. Japan lost a heavy cruiser, the U.S. lost a destroyer. Several Japanese ships were damaged, no other Americans were. Two small islands were taken in the North Pacific and one major island base was saved in the Central Pacific.

Discussion and comment about many of the customary evaluations and myths of the battle.

The Japanese plan began to go wrong early on.
• It is said that Yamamoto's plan was too complex; that forces were dispersed and unable to fight together. But we have seen that the entire Japanese fleet was engaged, it would have been foolish to put them all in one spot. The goals of the operation determined what ships were needed and where.

- The battleship antiaircraft guns could have been used to protect the carriers. But the biggest battleships followed to surprise the American fleet; their presence was successfully hidden from the Americans for the entire battle.
- Nugumo should have used some of his attack planes for scouting. This seems reasonable after the events, but the reconnaissance plan was layered. Nugumo did not know that each layer would fail in turn: 1) The flying boats were canceled when American ships blocked their fueling site. 2) The submarine patrols were two days late in arriving on station. 3) One scouting cruiser had catapult problems. 4) Weather would hide the American fleet. 5) Radio silence kept him without the latest news and unable to give orders.
- Nagumo's goal was to attack the carriers. He knew this and had kept half his attack bombers ready with torpedoes for any surface action that might appear and they did launch and hit *Yorktown*. The problem was that all of his Zero fighters had been engaged and needed a few minutes to land, refuel, and rearm. The American dive bomber attack occurred during an ideal five minute window. More fighters were needed. Americans were also to discover that the ratio of fighters to bombers had to be increased. Fighters had two roles: going to protect bombers making attacks, and staying to defend the carrier from attack. Yet at this time in the war, both sides had fighters equal to only a third the number of bombers carried.
- Was the Aleutians invasion a wasted effort? When seen after the fact, it was, but at the time, Japan needed to expand her early warning bases in all directions. Weather of the Aleutians was so bad that no useful scouting could be maintained, no attacks launched, were difficult to resupply, and the bases were later abandoned. But, while two carriers, two cruisers, and three destroyers from the Imperial Fleet, plus some screening destroyers for the convoy portion to the North Pacific, were used ; this caused Nimitz to send a task force from Pearl of 2 heavy and 3 light cruisers with 4 destroyers to join 9 DD already stationed in the north. These could have been added to the air defenses of the U.S. carriers at Midway. We cannot fault Yamamoto on this point because he weakened the U.S. defenses.
- Yamamoto followed traditional naval practice of leading the attack from a battleship at sea. In prior ages this put the commander in sight of the battle. In the age of aircraft, radio and, on the American side, radar, a battleship was a prison to the commander who could not see the whole battle by eye and he could not receive reports or communicate through radio silence. Nimitz stayed at Pearl Harbor near his radio receivers and transmitters which were not silenced. His senior commander, Fletcher, directed attacks at sea to destroy the four enemy carriers, losing his own in the process, then to Spruance that night who was, unfortunately, unable to find the enemy the next day or the next.

Of the American activities.

• Fletcher was commander of two carrier task forces. He sent Spruance to the attack while he recovered scouts. Although the Japanese carriers were under observation from Midway almost continually from when they exited the weather front, the carriers did not know the specific location nor the number of enemy carriers, and whether together or separate. Japan had ten carriers, six were big fleet carriers. Five were anticipated to attack Midway and only two were reported before Fletcher ordered the first American attack and only three were reported from that attack. Before and after that attack, a second carrier group was suspected and Fletcher held half of one carrier's aircraft in reserve and had scouts out to find those or other threats. Yet his *Yorktown* air group was so capably led, that when they launched an hour and a half after the spearhead by Spruance, they arrived on target at the same time and with fewer planes sank an enemy carrier.

• Training was a strategic issue ; like intelligence gathering or aircraft design, this was external to the fleet, yet greatly impacted the effectiveness of fleet action. *Hornet* was new to the fleet and while in Spruance's TF 16 was virtually worthless in the battle -- brave men died with little to show for it. *Enterprise* had lost Halsey to illnes; the inexperienced Spruance was not as effective as he would become later. Fletcher and *Yorktown* were forged in battle and they carried most of the weight of decisions and effect.

• Military intelligence success was all American, thanks to radio traffic analysis and the code breakers. The Japanese had a succession of failures -- flying boat, submarine, and scouting, and were hampered by radio silence.

• Technology was on the side of Japan. The Zero was the superb carrier fighter of the day. Japanese torpedoes ran straight and true. Their optics and night fighting ability were a threat, though not actually used in this battle. The Americans had radar to detect incoming attacks in time to position fighters in the direction of the incoming attack so as to allow a few additional minutes of intercept time -- radar then was nothing like what it became later in the war. American high altitude bombing by the Army Air Force was much anticipated and every miss that was close enough to splash water on a ship was reported as a sinking. The public was told that the victory at Midway was due to their bombers before the carriers returned with the real story. B-17s dropped 233 bombs into the Pacific without the impediment of steel ships. In fact, no level bombs and no torpedoes damaged the enemy -- only carrier-based dive bombers made hits. Submarines were ineffective on both sides. The Japanese subs were not of good caliber throughout the war. The American torpedoes -- air, surface, and submarine -- were almost worthless until fixed almost a year later.

- Myth that Spruance was the victor. Fletcher ordered Spruance to attack the two Japanese carriers that had been detected in a fair fight of 2 on 2, while Fletcher continued to search, itching to engage, the other two or three carriers with just his one carrier. The four carriers of the Japanese strike fleet were attacked under Fletcher's command. After his flagship was damaged, he released TF 16 to carry on the fight for the next day. Spruance was delayed by a submarine scare and the main force get away -- and they had stopped to refuel within range of his forces. But refueling stations on Midway had been destroyed and long-range scouts were seeking downed pilots. The third day he went after two damaged heavy cruisers and sank one of them. Later Spruance commanded the Fifth Fleet and his biographers looked to the first time he came to public attention and self-servingly credited him with the entire victory rather than the one cruiser sunk under his command.

The story of Midway is one of the most analyzed and written about of the Pacific War. Each side made their preparations. Luck and bravery were the final determinants. Add to the Annals of Heros : Waldron, McCluskey, Tomonaga and every torpedo plane pilot and crewman.
This summary only touches the highlights.

Who is Who ?
Japanese Admirals
Yamamoto -- Japanese chief of Combined Fleet, architect of Pearl Harbor and Midway.
Nagumo -- Japanese commander of Carrier Striking Forces, victor at Pearl Harbor, East Indies, and Indian Ocean.

American Admirals
Nimitz -- US Pacific Fleet Commander who determined to attempt the ambush.
Halsey -- US Commander of Pacific Carriers, hospitalized just before the Midway battle.
Fletcher -- Carrier Task Force Commander, first in history to stop a Japanese fleet, at the Coral Sea. He sank six enemy carriers in eight months ; the most successful admiral of the War.
Spruance – Carrier Task Group commander in *Enterprise* replacing the sick Halsey and was released to operate independently after the flagship, *Yorktown,* was torpedoed the second time. He sank a cruiser. After a period ashore to plan the next phase of the war, he returned to sea to command the Fifth Fleet.
Theobald -- Sent to intercept the invasion of the north, he stationed his ships to defend Alaska and the West Coast, missing the invasion in the Aleutians.

Japanese Carriers.

Akagi, Kaga, Hiryu, Soryu were big carriers that had led the Japanese victories from Pearl Harbor in the Eastern Pacific, to the East Indies in the Western Pacific, to Darwin, Australia, in the South West Pacific, to Ceylon in the Indian Ocean.

Shokaku and *Zuikaku* were normally with this group, but had been mauled by Fletcher at the Coral Sea and were refitting.
Zuikaku sailed to join the after-planned Aleutian ambush. Just over two years later, she ended as a sacrifice in the Philippines.

Junyo and *Ryujo* were medium and light carriers that attacked Dutch Harbor in the Aleutians. *Ryujo* will be put under by Fletcher at Eastern Solomons two months later, his sixth and final carrier kill.

American Carriers.

Yorktown (CV-5), Fletcher's flagship, had been damaged at the Coral Sea and rushed to Pearl for repair in time for the Midway action. *Yorktown* had scouting duty on the day of the battle and Fletcher sent Spruance with the other two carriers to attack while *Yorktown* recovered scouts, followed, sank a carrier, acted as air reserve and did scouting all at the same time and was attacked twice by air, taking three torpedo hits, and finally was attacked by submarine *I-68* with two more torpedo hits and went down after the battle.

Enterprise (CV-6), normally Halsey's flagship, an experienced ship, but with a staff that needed Halsey's tough hand. She became the most decorated ship of the Pacific War with the most combat engagements and survived the war.

Hornet (CV-8) was a new ship that had not yet seen battle but had carried Doolittle to raid Japan and had sailed to the South Pacific and returned for Midway. She was sunk a few months later in the South Pacific at Santa Cruz.

Saratoga (CV-3) was repaired from torpedo damage and rushed with replacement aircraft from San Diego to Pearl, then to the Midway fleet that was refueling for Alaska ; she provided 34 aircraft. She became Fletcher's new flagship and all returned to Pearl. "Sara" fought in the South Pacific, Indian Ocean, and Central Pacific ; she survived the war.

Battle of Midway -- June 4-6, 1942
In Greater Detail

The Story

The Japanese sent their entire fleet to capture Midway Island in the hopes the small U.S. Navy would come to fight and be destroyed. Nimitz got word of the plan and set an ambush. While the Japanese carriers were rearming after bombing Midway, Fletcher's airplanes attacked and destroyed three of four big carriers. Planes from the surviving carrier damaged *Yorktown.* A second U.S. strike put down the fourth enemy carrier. Damaged *Yorktown* was sunk by a submarine two days later.

The Cast

Admiral Earnest King - Chief of Naval Operations. Washington, D.C.
Admiral Chester Nimitz - CinC Pacific, Pearl Harbor.
Rear Admiral Frank Fletcher - Officer Tactical Command; Commander TF 17
Rear Admiral Raymond Spruance - Task Force 16, substituting for Halsey
    who was hospitalized.
Capt Elliott Buckmaster -- Yorktown (CV-5)
Capt George D. Murray -- Enterprise (CV-6)
Capt/RAdm Marc Mitscher -- Hornet (CV-8)
RAdm Thomas C. Kinkaid -- TF 16 Cruiser Group
RAdm William W. Smith -- TF 17 Cruiser Group
RAdm Robert A. Theobald -- TF 8 sent to Alaska

CARRIER STRIKING FORCES
United States Navy

RAdm Frank J. Fletcher -- Officer in Tactical Command, aboard *Yorktown*
    Chief of Staff – Capt. Spencer S. Lewis
Task Force 16 — RAdm Raymond A. Spruance
    Chief of Staff -- Capt Miles R. Browning
  Enterprise (CV-6) -- Capt George D. Murray
    Air Boss (CAG).- LCdr Clarence W McClusky
      VB-6 — 19 SBD-2,-3, Dauntless, Lt Richard S. Best *(11 lost)*
      VS-6 — 19 SBD-2,-3, Dauntless, Lt. Wilmer E. Gallaher *(9 lost)*
      VT-6 — 14 TBD-1, Devastator, LCdr Eugene E. Lindsey *(11 lost)*
      VF-6 — 27 F4F-4, Wildcat, Lt James S. Gray *(1 lost)*
  Hornet (CV-8)—Captain/RAdm Marc A. Mitscher
    Air Boss (CAG) - Cdr Stanhope C. Ring
      VB-8 — 19 SBD-2,-3, LCdr Robert R. Johnson *(5 lost)*
      VS-8 — 18 SBD-1,-2,-3, LCdr Walter F. Rodee
      VT-8 — 15 TBD-1, LCdr John C. Waldron *(15 lost) [all]*
      VF-8 — 27 F4F-4 LCdr Samuel G. Mitchell *(12 lost)*

Cruiser Group—RAdm Thomas C. Kinkaid (ComCruDiv 6)
  Minneapolis (CA-36)
  New Orleans (CA-32)
  Vincennes   (CA-44)
  Northampton (CA-26)
  Pensacola   (CA-24)
  Atlanta   (CLaa-51)
Destroyer Screen
  DesRon 1, Capt Alexander R. Early.
    Phelps (DD-360) [DL], Dewey (DD-349), Worden (DD-352),
    Aylwin (DD-355), Monaghan (DD-354)
  DesRon 6, Capt Edward P. Sauer
    Balch (DD-363) [DL], Conyngham (DD-371), Maury (DD-401),
    Benhan (DD-397), Ellet (DD-398), Monssen (DD-346).
Task Force 17 — RAdm Fletcher
  Yorktown (CV-5) — Capt Elliott Buckmaster
    Air Boss (CAG), LCdr Oscar Pederson
    VB-3 — 18 SBD-3, LCdr Maxwell F. Leslie *(12 lost)*
    VS-5 — 19 SBD-3, (really VB-5, renamed to avoid confusion)
      Lt Wallace C. Short, Jr
    VT-3 — 13 TBD-1, LCdr Lance E. Massey *(10 lost)*
    VF-3 — 25 F4F-4 (half from VF-42, experienced Yorktown pilots)
      LCdr John S Thach *(10 lost)*
  Cruiser Group, TG 17.2 — RAdm William W. "Peco" Smith
    Astoria (CA-34)(flag)
    Portland (CA-33)
  Destroyer Squadron, TG 17.4 — Capt Gilbert C. Hoover
    (ComDesRon 2)
    Morris (DD-417), Anderson (DD-411), **Hammann** (DD-412),
    Russell (DD-414), Hughes (DD-410), (one joined on June 5)
Submarines - RAdm Robert H. English, Commander, Submarine Force,
    Pacific Fleet, at Pearl Harbor.
    Cachalot, Flying Fish, Tambor, Trout, Grayling, Nautilus, Grouper,
    Dolphin, Gato, Cuttlefish, Gudgeon, Grenadier -- west of Midway ;
    Narwhal, Plunger, Trigger -- between Oahu and Midway ;
    Tarpon, Pike, Pinback, Growler -- north of Oahu.
Oiler Group
    Cimarron (AO-22) with Monssen (DD-436) stationed east of Midway.
    Platte (AO-24) with Dewey (DD-349) with TF 17.

From Midway Island.
Marine
 20 F2A-3 Buffalo (*13 lost*) ;   7 F4F-3 Wildcat (*2 lost*)
 11 SB2U Vindicator (*4 lost*) ; 16 SBD-2 Dauntless (*8 lost*)
Army Air Force
 4 B-26 Marauder (*3 lost*) ;   19 B-17 Flying Fortress (*2 lost*)
Navy
 32 PBY-5,-5a Catalinas (*1 lost*)
  6 TBF Avenger (a new type, *5 lost*)

IMPERIAL JAPANESE NAVY

COMBINED FLEET — Adm Isoroky Yamamoto, in *Yamato*
 ADVANCE EXPEDITIONARY FORCE — VAdm Teruhisa
  Komatsu
  16 submarines - 10 for Midway, 6 for Alaska
 FIRST MOBILE FORCE — VAdm Chuichi Nagumo
  Akagi , Kaga , Hiryu , Soryu (CV) - fleet carriers
  Haruna (BB) , Kirishima (BB) - battleships
  Tone (CA) , Chikuma (CA) – scouting heavy cruisers
  12 DD – destroyers
  6 AO – oilers ; plus other ships
 MIDWAY OCCUPATION FORCE — VAdm Nobutake Kondo
  2 BB ; 4 CA ; 7 DD ; 6 AO ; Nuiho (CVL) ; other ships
  CLOSE SUPPORT GROUP — RAdm Takeo Kurita
   Kumano (CA), Suzuya (CA), **Mikuma** (CA), *Mogami* (CA)
   Asashio (DD), Arashio (DD) ; Nichiei (AO)
  TRANSPORT GROUP — RAdm Raizo Tanaka
   12 AP/AK ; 1 CL ; 10 DD ; 1 AO ; other ships
  MINESWEEPING GROUP
   4 minesweepers; 3 PC; 2 AK
 MAIN BODY (FIRST FLEET) — Adm Yamamoto
  Yamato, Mutsu, Nagato -- battleships ; 13 DD ;
  Hosho (CVL) ; 2 seaplane carriers – Chiyoda, Nisshin
 GUARD FORCE — VAdm Shiro Takasu
  Hyuga, Ise, Fuso, Yamashiro – battleships
  Kitakami, Oi (CL) ; 11 DD ; 4 AO.
  Stationed between Midway and Alaska able to go either way.
 SUBMARINE FORCE
  10 I-boats (3 carrying aviation gasoline)
 SHORE-BASED AIR FORCE
  Wake, Kwajalein, Aur, Wotje – 10 bombers, 72 torpedo planes,
    72 Zeros, 24 flying boats.

SIDESHOWS
Alaska United States Navy

Task Force 8 — RAdm Robert A. Theobald in *Nashville*.
  Main Body.
    Nashville   (CL-43) flag
    Indianapolis (CA-35)
    Louisville   (CA-28)
    St. Louis   (CL-49)
    Honolulu   (CL-48)
  DesDiv 11 : Gridley (DD-380), McCall (DD-400), Gilmer (DD-233), Humphreys (DD-236).
  TG 8.1 Air Search Group.
    Williamson (AVD-2), Gillis (AVD-12), Casco (AVP-12) with 20 PBY of Patrol Wing 4 ; 1 B-17.
  TG 8.2 Surface Search or Scouting Group
    Charleston (PG) ; Oriole (AO) ; 14 YP ; 5 Coast Guard cutters.
  TG 8.3 Air Striking Group — Brig Gen William O. Butler, USA.
    Ft. Randall, Cold Bay — 21 fighters, 14 bombers.
    Ft. Glenn, Umnak — 12 fighters.
    Kodiak Naval Station — 32 fighters, 5 bombers, 2 light bombers.
    Anchorage — 44 fighters, 24 bombers, 2 light bombers.
  TG 8.4 Destroyer Striking Group — 9 DD
    Case (DD-370), Reid (DD369), Sands (DD-243), King (DD-242), Kane (DD-235), Brookes (DD-232), Dent (DD-116), Waters (DD-115), Talbot (DD-114).
  TG 8.5 Submarine Group - 6 older 'S' boats
    S-18, S-23, S-27, S-28, S-34, S-35.
  TG 8.9 Tanker Group — Sabine (AO), Brazos (AO), SS Comet.

Aleutians Japanese Navy

  NORTHERN AREA FORCE - VAdm Hosogaya
    Nachi (CA) ; 2 DD ; 2 AO ; 3 AK
  Second Mobile Force - RAdm Kakuji Kakuta
    Junyo (CV) ; Ryujo (CVL) -- carriers
    Maya (CA) , Takao (CA) ; 3 DD ; 1 AO
  Occupation Force Adak-Attu
    Abukuma (CL) ; 5 DD ; 2 AP ; Kimikawa (AV) ; plus minesweepers
  Occupation Force Kiska
    Kiso (CL), Tama (CL) ; Asaka (CM) ; 3 DD ; 2 AP ; 3 SC
  Submarines: 6 I-boats

Till they met Fletcher

U.S. West Coast
TF 1 (Battleships)   VAdm William S. Pye.
Pye had his battleships sunk at Pearl and commanded the replacements that were returned from the Atlantic or as repaired. The BB fleet was based in San Pedro (Los Angles). There was not room, insufficient fuel, tankers, or need for slow BBs in Pearl at this time. Not included in the Midway planning, he sortied to provide a backstop in case the Japanese made a foray towards the U.S. mainland or if they broke through from Midway to raid Hawaii. Although a respected naval leader before Pearl Harbor, his career withered.
31May42. BatDiv 4 sortied, RAdm Walter S. Anderson
   Maryland (BB-46), Colorado (BB-45) ; 2 DD, [a 3rd joined next day]
5June42 . sortied :
   Mississippi (BB-41), Idaho (BB-42) ; New Mexico (BB-40),
   Tennessee (BB-43) , Pennsylvania (BB-38) ; 5 DD ; and
   Long Island (AVG-1) *[later known as CVE-1. Test as combat scout.]*

Battle Losses at Midway
U.S.N.
   Yorktown (CV-5) , Hammann (DD-412)
   144 aircraft ; 104 pilots and aircrew ; 258 sailors
      This does not include Midway based losses : possibly 70 pilots and aircrews.
Japanese
   Fleet carriers -- Akagi, Kaga, Soryu, Hiryu
   Heavy cruiser -- Mikuma
   250 aircraft ; about 110 pilots and aircrew ; almost 3,000 sailors

PEOPLE:
King . Member of the Joint Chiefs, attended meeting with military and civilian leaders to lobby for increased support for the Pacific War (Allied policy was "Germany first"). Interfered with Nimitz for the first six months. His D.C. staff and the Pacific staff frequently differed on the interpretation of Japanese intentions. King was concerned with an attack on West Coast, favored battleships, and wanted the carrier escort of cruisers and destroyers to make surface attacks. He represented the political desires from D.C.

Nimitz , and his admirals won the Pacific War, though MacArthur was given command of defeated Japan. Nimitz had to convince King the next battle would be for Midway and took the giant risk with the small U.S. fleet to try to ambush the much larger Japanese force.

Fletcher . Officer in Tactical Command, fresh from Battle of the Coral Sea four weeks before where he had traded carriers, in the first carrier battle in history, with the Imperial Japanese Navy which caused them to withdraw their invasion force. He had sunk one and damaged two carriers that did not make it to Midway. Six weeks later he would save Guadalcanal in the Battle of

Eastern Solomons. King, over Nimitz's objection, sent Fletcher to quell the public fears of invasion through Alaska. Fletcher had sunk six enemy carriers, a record unmatched. Later he became Commander North Pacific and accepted surrender of the Japanese northern fleet forces.

Spruance . Substituted for Halsey (in hospital) as TF 16. Had already been picked for Nimitz' next chief of staff. Later commanded U.S. Fifth Fleet.

Halsey . The senior carrier task force commander was in the hospital and was out of the war for the period June to September 1942. Commanded South Pacific for the later Solomons campaign and the Third Fleet in attacking Japan

At this stage in the war, captains assigned to fleet carriers had already been selected for flag rank : Mitscher, Murray, Noyes.

Mitscher . *Hornet,* had already made admiral in May, yet remained as captain of *Hornet,* but he performed so poorly he was sent ashore for 18 months. Came back later to command the carrier force at the end of the war.

Murray . *Enterprise,* was actually on his way out, but performed so well at Midway he did make Rear Admiral. Given TF 17, he aggressively moved through the South Pacific until *Hornet* was sunk at Santa Cruz. Later became ComAirPac.

Buckmaster . *Yorktown* was the most successful captain at Midway, but because *Yorktown* did not rollover, though her high flight deck was touching the water when he ordered the ship abandoned, his career suffered stagnation.

Ramsey . *Saratoga,* became RAdm, Task Group commander in New Georgia campaign which included *Sara.*

Other Admirals

Kinkaid . A cruiser admiral, received a carrier task force replacing Fletcher. Fought the Battle of Santa Cruz which turned back the enemy while losing carrier *Hornet.* Sent to replace Theobald in Alaska and retake islands. Came back in Nov'43 to command MacArthur's Navy. Named commander Central Philippines Attack Force.

Theobald . Not privy to Nimitz's secret information, he positioned his TF too far east. He moved up to command Alaska Frontier, soon replaced by Kinkaid, who was later replaced by Fletcher who became commander all North Pacific Theater.

Fitch . TF 11, *Saratoga.* Had been Fletcher's air commander in *Lexington* at the Coral Sea. He arrived too late for Midway action, but *Saratoga* provided replacement aircraft. Made ferry cruise to Midway, but was too valuable to naval air administration to continue in sea duty. To Commander Naval Air Forces South Pacific.

Smith . Previously Chief of Staff, Pacific. Cruiser admiral, moved up to command TF 8 in Alaska. Held a succession of senior positions. VAdm, Service Force, March 1945, for Okinawa.

Other Officers

Browning – TF 16 Chief of Staff provided advice to inexperienced Spruance. Credited with much of victory, but made many mistakes that cost planes. Bright, but temperamental, sent ashore until Halsey returned him to Chief of Staff, South Pacific, had brief command of Hornet II, then permanently assigned to the Staff College.

CARRIERS

Lexington (CV-2) was lost at the Coral Sea one month earlier while successfully repulsing he Japanese invasion of Port Moresby in company with *Yorktown*.

Saratoga (CV-3) had completed repairs at Bremerton, WA, from torpedo damage. Hastily departed San Diego 1June42 under Captain Dewitt C. Ramsey without Task Force 11 Commander, RAdm Aubrey W. Fitch. Fitch in *Chester* (CA-27) arrived San Diego the next day, but had to provision and departed 4 June, arrived Pearl 8 June after *Sara* had left again. *Saratoga, San Diego* (CLaa-53), and 4 DD had arrived Pearl 6June, departed north on 7June with a surplus of aircraft to replenish *Enterprise* and *Hornet*. Joined Fletcher 8June becoming his flagship of TF 17. *Sara* returned Pearl, became TF 11 (Fitch), departed 18June with replacement aircraft for Midway.

Ranger (CV-4) was an experiment with smaller, more economical carrier design and was principally an aircraft ferry to Africa and served in the Atlantic during the war.

Yorktown (CV-5) returned to Pacific from Neutrality Patrol on 1Jan42 to become Fletcher's flagship of TF 17. Escorted troops to Samoa and attacked Gilberts on the way back. Attacked invasion at Salamaua-Lae. *Yorktown* was damaged at the Coral Sea, hastily repaired at Pearl in time to fight at Midway. Damaged again, she was finally sunk by a submarine while under tow.

USS Yorktown after Midway.
Fletcher's score is 4 : 1

Enterprise (CV-6) Sister ship to *Yorktown*. The "Big E" was the most decorated ship of WWII, fought in 20 battles. Enough said !

Wasp (CV-7) was smaller, build from lessons learned with *Ranger*. and was in the Atlantic, twice ferrying Spitfires to Malta. She was sent to replace *Lexington* in the Pacific arriving in mid-June, after the Battle of Midway . She was sunk by submarine attack 3 months later while escorting troops to reinforce Guadalcanal.

Hornet (CV-8) as a brand new ship on shakedown was tagged for Doolittle's B-25 raid on Tokyo and sent to the Pacific. She was lost at Santa Cruz (Guadalcanal) four and a half months after Midway.

Long Island (AVG-1) was converted from a merchant hull in 68 days, June'41, as an auxiliary to free fleet carriers from ferry duty. She was being tested as a scout with TF 1, the battleship fleet, during Midway. Three months later *Long Island*, as ACV-1, delivered the first planes to Guadalcanal. This class of ship was redesignated from auxiliary to combatant, CVE, on 15July43. Kaiser built 50 CVE, escort carriers, from merchant hulls with flight decks, in a 12 month period for ferry, anti-sub, and occasionally, attack combat roles. They commissioned July'43 to mid'44 for service in the later war.

EQUIPMENT.

F4F-4 Wildcat had folding wings and 6- .50 cal machine guns. Its flight performance against a Zero was no better than the fixed wing, 4-gun, F4F-3 Wildcat, but more planes could be carried.

History. Grumman F4F-1 was a biplane design, withdrawn, and a single wing version, F4F-2, submitted that lost competitive bidding to the F2A Buffalo in 1938 The F4F-3 was fitted with a larger engine and supercharger and the Navy ordered 78 in Aug'39. Early production went to the British Fleet in July'40 and to the USN in Dec'40. Ninety-five more aircraft were ordered for 1941. The F4F-4 had 2 more guns, folding wings, and self-sealing gas tanks with delivery starting about the time of Pearl Harbor. A thousand were built as FM-1 by General Motors.

PBY-5 Catalina Flying boat with two engines mounted on a high wing. The **-5a** had retractable wheels for sea or land use. This added weight, reducing range, but gave more flexibility.

1.1" antiaircraft gun. Usually a quad mount; prone to jamming, these were being replaced with the **40mm (1.5") Bofors** gun in dual and quad mounts. **.50 cal machine gun** replacement had started before the war with the **Oerlikon 20mm** (.78") for close-in defense and 20mm were eventually mounted everywhere there was room aboard ship.

BATTLE OF MIDWAY TIMELINE

MAY 26.
- *Enterprise* and *Hornet* arrive Pearl Harbor from South Pacific.
- Halsey to hospital, Spruance is replacement.
- Japanese submarines to refuel seaplanes find U.S. seaplane tender at French Frigate Shoals.
- Japanese Aleutians carriers underway from Ominato.

pm   Japanese Midway carrier strike force leaves Tankan Bay.

MAY 27.
am   Japanese amphibious forces leave Mutsukai Bay for Aleutians.
1400   *Yorktown* arrives Pearl Harbor, straight into drydock.
pm   Liberty parties ordered back to ship by 1830.
pm   Japanese Occupation force departs Saipan.
pm   Japanese Close Support group (bombardment force) departs Guam.

MAY 28.
am   *Enterprise, Hornet* TF 16 (Spruance) depart Pearl Harbor.
1800   Air group flies to carriers.

MAY 29.
pm   *Yorktown* leaves drydock, reprovisions.

MAY 30.
0900   *Yorktown* TF 17 (Fletcher) departs.
- Weather turns bad as TF 16 nears Midway.
- TF 16 arrives 350 miles NE of Midway.

MAY 31.
- TF 16 refuels.
- Japanese spy network seems to have been shut down by this date.

JUNE 1
- *Saratoga* departs San Diego.
- Bad weather in the whole Midway area.
- Japanese enter storm front that provides protection from Midway.
- TF 17 refuels.

JUNE 2.
1600   TF 16 and TF 17 meet up 325 miles NE of Midway.

JUNE 3.
0807   Dutch Harbor attacked by 12 aircraft from light carrier *Ryujo*.
        Junyo flights turned back by weather.
0900   PBY from Midway sights Japanese transport force 700 miles west.
        Japanese submarine line established North of Hawaii. U.S. carriers had already passed.
1624   Nine B-17s attack transport force three times. No hits.

JUNE 4.

0143	PBYs drop three torpedoes on transport force, one hit an oiler, it survives.
0400	Japanese start launch of 108 aircraft attack towards Midway: 36 Kate, 36 Val, 36 Zeke (Zero).
0431	Ten SBD launched by *Yorktown* to search 100 miles north and west. *Tone's* catapult malfunctions, delays scout assigned to the sector with U.S. forces.
0545	"Many planes heading Midway" report by PBY.
0607	Fletcher orders Spruance "Proceed southwesterly and attack carriers as soon as definitely located."
0620	Midway fighters attack Japanese aircraft.
0622	"Two carriers, one battleship" report is in error by 25 miles.
0630	Bombs drop on Midway.
0630	*Yorktown* recovers scout planes and proceeds SW.
0700	Japanese bombers finish attack on Midway, call for a second strike.
0700	Spruance launches aircraft from *Enterprise* and *Hornet* without knowing thelocation of enemy.
0710	Four Army B-26 bombers carrying torpedoes and six of the new Navy TBF torpedo bombers from Midway attack *Akagi*.
0715	Nagumo orders to rearm torpedo planes with bombs for second attack on Midway.
0728	Nagumo receives report of ten U.S. ships.
0745	Nagumo stops unloading torpedoes.
0745	*Enterprise* bombers sent ahead, fighters and torpedo planes never join up.
0800	All TF 16 attack groups are in air. Where-to is various. *Hornet* groups headed far north of target area. Waldron's torpedo group breaks off towards the target.
0800	Sixteen Marine SBD from Midway attack *Hiryu*, 8 SBD destroyed, no hits.
0815	B-17s from Midway attack, no hits.
0820	Eleven Marine SB2U Vindicators attack battleship *Haruna*, no hits, two SB2U lost.
0820	Nagumo receives report of one U.S. carrier, reinstalls torpedoes.
0820	Japanese Midway strike returns.
0837	Nagumo starts recovery of strike aircraft.
0838	*Yorktown* launches half of attack craft, retains reserve for other, unlocated carriers.
0855	Nagumo warned by scout "Sight ten enemy attack planes heading towards you. (Waldron,VT-8)
0910	*Enterprise* fighters find fourth Japanese carrier. U.S. carriers never hear of location.
0915	VF-8 turns back. Unable to find *Hornet*, all ten ditch together.
0918	Japanese attack aircraft recovered.
0918	Nagumo turns northeast to close with American fleet.

Till they met Fletcher *119*

0920	Waldron's VT-8 sighted by Japanese screen.
0930	VT-6 finds Japanese. 120 mph torpedo bombers chase 40 mph ships.
0935	McClusky sights Japanese DD heading towards its fleet.
0950	VT-8 attacks, all shot down.
0955	VT-6 attacks, 9 splashed, 5 got away.
0956	VF-6 radio report gets through.
1000	*Enterprise* fighters leave short on fuel.
1002	McClusky VB-6 SBD sight carriers.
1003	*Yorktown* VB-3 SBD sight carriers
	Enterprise SBDs had reached the target area, but finding no enemy had to search for over an hour.
	Hornet's SBDs headed far north of target area, came south too far west; most landed on Midway, 3 ditched for lack of fuel.
	Yorktown's SBDs, launched hour and 20 minutes after TF 16's, travels directly to target arriving at same time.
1010	*Yorktown* starts a coordinated attack.
1020	*Enterprise* SBDs : VS-6 on *Kaga* (4- 500# hits) and VB-6 (2- 1000# hits) on *Akagi*.
1022	*Yorktown* VB-6 SBDs dive on *Soryu* (3- 1000# hits).
1050	*Hiryu* launches attack for *Yorktown*, 18 dive bombers, 6 Zeros.
1120	*Yorktown* launches 10 scouts to 200 miles.
1156	Nagumo orders surface fleet to attack American carrier fleet.
1210	*Yorktown* takes three bomb hits.
1331	*Hiryu* launches attack #2 on *Yorktown* ; 10 KateTB, 6 Zeros.
1400	Japanese heavy cruiser bombardment force races ahead.
1430	*Yorktown* back to speed.
1444	*Yorktown* torpedoed twice by second air attack.
1445	*Yorktown* scouts find the fourth carrier.
1455	*Yorktown* abandons ship.
1500	*Hornet* ready to launch, but not told of *Enterprise* plans.
1510	*Hornet* recovers 11 SBD that had diverted to Midway.
1528	*Enterprise* launch 25 SBDs without fighter escort -- 11 from *Enterprise* and 14 from *Yorktown*.
1558	*Hornet* launches 16 SBD.
1600	*Ryujo* and *Junyo* had launched 26 aircraft on Dutch Harbor ; Yamamoto orders Aleutian carriers south to Midway.
1610	**Soryu** sinks.
1704	*Hiryu* attacked by SBDs, four hits.
1720	*Hornet* bombers attack screen, no hits.
1811	Fletcher releases TF 16 while sailing east with survivors.
1820	*Enterprise* recover SBDs and CAP.
1857	*Hornet* recovers SBDs and CAP.
1915	U.S. ships turn east for the night.
1925	**Kaga** sinks.
1930	Yamamoto orders Northern Force to return and occupy western Aleutians.

JUNE 5.
0020   Yamamoto cancels cruiser bombardment of Midway.
0255   Yamamoto orders a general withdrawal.
0342   Heavy cruisers *Mogami* and *Mikuma are* damaged in collision while avoiding submarine *Tambor* (SS-198).
0500   **Akagi** scuttled.
0825   Bad weather, *Enterprise* launches CAP.
         Midway bombers attack damaged cruisers to no effect.
0900   **Hiryu** sinks.
1100   TF 16 heads NW to search for reported Japanese carrier.
1200   Brown wanted to make long-range attack, overruled by Spruance.
1308   Fleet tug *Vireo* (AM-52) begins towing *Yorktown* towards Pearl Harbor.
1512   *Hornet* launches 12 SBD to the reported area, nothing found.
1530   *Enterprise* launches 30 SBD, ditto
1543   *Hornet* launches 14 more SBDs, ditto
1800   Salvage crew to *Hammann* and sent towards *Yorktown*.
2040   Spruance gives up the chase and turns to try to find cruisers
2320   Yamamoto sends BBs *Kondo* & *Hiei*, CVL *Zuiho*, CAs *Tone* & *Chikuma* and 14 *subs* to reinforce Japanese northern force and to ambush U.S.

JUNE 6.
a.m.   *Yorktown* crew taken on by destroyers transferred to cruiser *Portland*.
0440   Salvage crew arrives on *Yorktown*.
0440   *Portland* with 2000 survivors heads for *Fulton* (AS-11) sent from Pearl.
0500   *Enterprise*, 350 miles NW Midway launches 18 SBD to search 200 miles out. Finds the two damaged cruisers with two destroyers. *Mogami* had bow sheared off, *Mikuma* now appears larger and is reported as a battleship.
a.m.   *Saratoga* arrives Pearl Harbor from San Diego.
0757   *Hornet* launches 24 planes to attack cruisers.
0815   *Enterprise* recovers scouts.
0930   *Hornet* squadrons attack cruisers, make six hits.
1035   *Hornet* recovers attack planes
1045   *Enterprise* launches 31 SBD, 12 F4F, 3 TBD.
1245   *Enterprise* air groups attack cruisers.
1334   *Yorktown* and **Hammann** torpedoed. Destroyer sinks immediately
1400   *Hornet* launches a second attack on cruisers with 24 SBD.
1415   *Enterprise* recovers air groups.
1500   *Hornet* fliers make six 1000- pound bomb hits on *Mikuma*, one on *Mogami*.
1550   *Yorktown*, "Abandon ship".
1550   Yamamoto steams south hoping to attack Spruance.
1907   TF 16 lands CAP, turns east to refuel. Battle over.
Sunset   **Mikuma** sinks; *Mogami* and 2 DD escape.

JUNE 7.
0315 Japanese troops occupy Attu and Kiska without opposition.
0501 **Yorktown** goes under.
0700 Yamamoto turns back to Japan.
am   *Saratoga* sets sail with replenishment aircraft, 107 planes on board.
      In the wake of the battle, efforts to locate downed aviators persist over the ensuing days. 27 aviators rescued during ten days after the battle.

JUNE 8.
      U.S. and Japanese each refuel in bad weather.
1112 *Saratoga* arrives. Fletcher moves to *Saratoga*, new flag of TF 17.
      Five VF-8 pilots recovered by PBY; three more next day.

JUNE 9. No flying from bad weather

JUNE 10.
      TF 16 and TF 17 link up in morning fog.
      Wasp (CV-7), *North Carolina* (BB-55), *Quincy* (CA-39),
      San Juan (CL-54), and 6 DD transit the Panama Canal to become
      TF 18 (RAdm Noyes).

JUNE 11.
am   *Saratoga* transfers replacement aircraft -- 10 SBD, 5 TBD for
      Enterprise ; 9 SBD, 10 TBF for *Hornet*.
1100 *Saratoga* (with inexperienced crew) returns to Pearl to report.
1100 TF 16 turns north to Alaska.
1300 Nimitz recalls TF 16 to Pearl Harbor.

JUNE 12.
      Zuikaku and *Zuiho,* 2 BB, 4 CA, DDs arrive Aleutians to reinforce
      Japanese northern ambush force ; now with four carriers.
      Saratoga launches training flights.
      Army Air Force declares victory by B-17s.

JUNE 13.
      Three surviving carriers arrive Pearl Harbor.
      The first part of the Pacific War is over.

U. S. CARRIER AIRCRAFT at BATTLE of MIDWAY

Aircraft Type	Aboard	Launched to attack	Engaged Enemy	Lost (include ditchings)
F4F	79	28	8	16
SBD	101	67	37	24
TBD	41	41	41	37
Total	221	116	86	80

Others were lost in defense of *Yorktown* and in the attack on cruisers.

Midway Atoll, Nov 1941, was a peaceful airfield mid-way between Hawaii and the Philippines with hotel for transpacific passengers. Eastern Island is in foreground, Sand Island beyond.

Admiral Chester Nimitz, 1942. He bet it all. And won the Pacific War.

Japanese Carriers of World War II.
Ten Carriers at the time of Pearl Harbor

The First Year

Name	Type	Dec'41	Jan'42	Feb	Mar	Apr	May	June
Akagi	CV	PH	Rabaul	Truk	Java	Ceylon	home	*Midway*
Kaga	CV	PH	Rabaul	Truk	Java	repair	home	*Midway*
Soryu	CV	PH,WI	Molucca	Darwin	Java	Ceylon	home	*Midway*
Hiryu	CV	PH,WI	Molucca	Darwin	Java	Ceylon	home	*Midway*
Shokaku	CV	PH	Rabaul	home	home	Ceylon	CoralSea	repair
Zuikaku	CV	PH	Rabaul	home	home	Ceylon	CoralSea	home, Kiska
Hosho	CVL	Pilot Training	→	→	→	→		Midway
Ryujo	CVL	P.I.	Java	Java	Java	Bengal	Mitsu	DutchHarbor
Zuiho	CVL	P.I.	Java	Java	Java	Kure	Kure	Midway
Taiyo	CVE	P.I.	Java	ferry	→	→	Kure	repair
Shoho	CVL		*26Jan*	ferry	→	training	***Coral Sea***	
Junyo	CV					comm	*05May*	DutchHarbor
Unyo	CVE					comm	*31May*	
Hiyo	CV							
Chuyo	CVE							
Ryuho	CVL							

Japanese Heavy Aircraft Carrier KAGA
38,000 tons; 812 feet; 28 knots; 72 (+18) aircraft

Photo # NH 73060 Japanese aircraft carrier Kaga, after her 1936 modernization

Name	Type a/c	July	Aug	Sep	Oct	Nov	Dec'42
Akagi	CV 63	Sunk at Midway					
Kaga	CV 72	Sunk at Midway					
Soryu	CV 63	Sunk at Midway					
Hiryu	CV 63	Sunk at Midway					
Shokaku	CV 72	Japan	E.Sol	Truk	SanCr	Japan	Japan
Zuikaku	CV 72	Kiska	E.Sol	Truk	SanCr	Japan	Japan
Hosho	CVL 30	training	→	→	→	→	→
Ryujo	CVL 31	Kiska	***E.Sol***	Sunk at Eastern Solomons			
Zuiho	CVL 24	Kiska	Japan	ferry	SanCr	GuadalC	Japan
Taiyo	CVE 30	Japan	Truk	Truk	Japan	ferry	ferry
Shoho	**CVL 30**	Sunk at Coral Sea					
Junyo	CV 45	Kiska	Japan	Japan	SanCr	GuadalC	Truk
Unyo	CVE 30	ferry	→	→	→	→	→
Hiyo	CV 45	*31July*	Japan	Japan	GuadC	Truk	Japan
Chuyo	CVE 30				comm	*25Nov*	ferry
Ryuho	CVL 31				comm	*28Nov*	damaged

Key: ***Bold italic*** = sunk,
Italic = commissioned
ferry = mostly from Yokosuka to Truk with aircraft for replacement of losses in the Solomons

From the time of the naval battles around Guadalcanal in November 1942 until the Marianas campaign in June 1944, there were no fleet to fleet engagements of consequence. During this period, Japan produced a new big carrier, *Taiho*, and converted two seaplane tenders and a submarine tender to light carriers, three passenger liners to escort carriers, two battleships to handle aircraft, and neared completion of the world's largest aircraft carrier, *Shinano*. In the same period, the United States commissioned 10 Essex class, large fleet carriers ; 9 Independence class light carriers ; and 39 Casablanca class escort carriers (of 50 produced by Kaiser), though many of these served in the Atlantic.

For Operation Olympus, invasion of the main island of Honsho in the spring of 1946, the U.S. would have had over 100 carriers of all sizes from USN and RN, Pacific and Atlantic fleets. Japan had built or converted an additional ten carriers and had four under construction at war end.

Till they met Fletcher

Fleet Seaplane Carriers, 1942

Name	Tons K	Dec	Jan	Feb	Mar	Apr	May
Kamoi	/AO 19.2	Truk	Rabaul	Rabaul	Rabaul	→	→
Chitose	CVS 11.4	P.I.	P.I.	Celebes	Java	New Gui	Kure
Chiyoda	CVS 10.9	Kure	Kure	Celebes	Kure	Kure	Truk
Kamikawa	6.8	Malaya	Indochina	Java	Rabaul	Solomon	Tulagi
Kimikawa	6.8	Kuriles	Kuriles	Kuriles	Kuriles	Kuriles	Kiski
Kiyokawa	6.8	Wake	Rabaul	NewGuinea	Lea	Japan	Solomons
Kunikawa	6.8	transport	~	~	~	~	~
Mizuho	CVS 10.9	P.I.	Molucca	MakassarS	Java	Japan	*sunk*
Sagara	7.2	Malaya	Indochina	EastIndies	EastIndie	Burma	EastIndies
Sanuki	7.2	P.I.	Borneo	EastIndies	P.I.	P.I.	P.I.
San'yo	8.4	Malaya,PI	Borneo	Indochina	Java	Java	Indochina
Nisshin	CVS 11.3	~	~	*commission*	Kure	East Ind	Kure
Akitsushima	AV 4.6	~	~	~	~	*commiss*	Rabaul

Name	June	July	Aug	Sep	Oct	Nov	Dec
Kamoi	Rabaul	→	→	→	EastIndie	EastIndies	EastIndie
Chitose	Midway	Japan	E.Solom	Truk	GuadalC	Sasebo	Sasebo
Chiyoda	Midw(ms)	Kiska	CVL	~	~	~	~
Kamikawa	Atti	Kiska	E.Solom	Solomon	Solomon	Solomons	Japan
Kimikawa	Kiska	Japan	Kiski	Kiski	Kiski	Kiski	Kiski
Kiyokawa	Truk	NewGuinea	Rabaul	Rabaul	Rabaul	NewGuinea	AK
Kunikawa	Japan	*commiss*	GuadalC	Solomon	Solomon	Japan	Solomons
Mizuho	*sunk May*						
Sagara	Singapore	Malaya	Malaya	Malaya	Malaya	Malaya	AK
Sanuki	P.I.	Japan	GuadalC	Solomon	Rabaul	Rabaul	AK
San'yo	Japan	EastIndies	GuadalC	GuadalC	GuadalC	Solomons	*damaged*
Nisshin	Midw(ms)	Kure	Kure	Rabaul	GuadalC	Solomons	
Akitsushima	Rabaul	Rabaul	GuadalC	Solomon			

Key: Tons K – displacement in thousands of tons
P.I.   – Philippine Islands
MidW – battle of Midway
ms   – mini-subs
AK   – started conversion to transport
~   – before or after conversion
→   – continue in same function

Chronology of Japanese Carriers and Seaplane Carriers.

1941

6 Dec. Yacht *Isabel* (PY-10) ordered by FDR as picket, is sighted by a floatplane from Japanese seaplane carrier *Kamikawa Maru* off IndoChina. Later in the day, *Isabel* is recalled from the suicidal mission.

8 Dec. Seaplane tender (destroyer) *William B. Preston* (AVD-7) is attacked by fighters and attack planes from Japanese carrier *Ryujo* in Davao Gulf, P.I.

14Dec. Submarine *Seawolf* (SS-197) torpedoes the Japanese seaplane carrier *San'yo Maru* off Aparri, P.I.; one torpedo hits the ship but does not explode.

16Dec. Japanese Pearl Harbor Attack Force detaches carriers *Hiryu* and *Soryu*, heavy cruisers *Tone* and *Chikuma*, and two destroyers to reinforce a second attack on Wake Island.

21Dec. Planes from carriers *Soryu* and *Hiryu* bomb Wake Island.

22Dec. Japanese bombers and attack planes, covered by fighters from carriers *Soryu* and *Hiryu*, bomb Wake Island.

23Dec41. Planes from carriers *Hiryu* and *Soryu*, as well as seaplane carrier *Kiyokawa Maru* provide close air support for the invasion of Wake Island.

1942.

27Jan42. USAAF B-17s bomb and damage Japanese seaplane carrier *Sanuki Maru* off Balikpapan, Borneo.

10Feb42. USAAF LB-30s bomb and damage Japanese seaplane carrier *Chitose* in Makassar Strait south of Celebes.

15Feb42. ABDA striking force (RAdm Doorman, RNN) is attacked by Japanese naval land attack planes as well as carrier attack planes from *Ryujo*.

1Mar42. Japanese heavy cruisers *Myoko, Ashigara, Haguro,* and *Nachi* engage three Allied ships fleeing Java, sinking British heavy cruiser HMS **Exeter**, (of *Graf Spee* fame) and destroyer HMS **Encounter**. Destroyer *Pope* (DD-225), escapes the cruisers but is located and bombed by floatplanes from seaplane carriers *Chitose* and *Mizuho*. *Pope* is then located by carrier attack planes from *Ryujo* and bombed. Scuttling is in progress when *Myoko* and *Ashigara* deliver the *coup de grace* with gunfire.

10Mar42. Two U.S. carriers attack the Japanese invasion fleet off Lae and Salamaua, New Guinea, sinking three and damaging ten ships including seaplane carrier *Kiyokawa Maru*.

6Apr42. Indian Ocean, Central Group, formed around carrier *Ryujo* attacks shipping, while *Akagi, Soryu, Hiryu, Shokaku, Zuikaku* attack British fleet at Ceylon.

9Apr42. **PT-34** is bombed and strafed by floatplanes from Japanese seaplane carrier *Sanuki Maru* and destroyed off Cauit Island, P.I.

2May42. Submarine *Drum* (SS-228) torpedoes and sinks Japanese seaplane carrier **Mizuho** off the south coast of Honshu.

7May42. Battle of the Coral Sea. A light carrier from the invasion support force, **Shoho**, is sunk. "Scratch one flattop."

8May42. **Battle of the Coral Sea** concludes as Japanese Carrier Strike Force formed around carriers *Shokaku* and *Zuikaku* is located and taken under air attack. SBDs from *Lexington* (CV-2) and *Yorktown* (CV-5) damage *Shokaku* and force her retirement. *Zuikaku*'s air group suffers heavy losses. Damage to *Shokaku*, as well as to *Zuikaku*'s air group, prevents the use of those two carriers for several months, thus making them unavailable for the Battle of Midway.

27May42. Navy Day ceremony Inland Sea: *Akagi*, *Kaga*, *Soryu*, and *Hiryu*.

3June42. As part of the overall Midway plan, Japanese Second Strike Force bombs Dutch Harbor, Alaska, with planes from carriers *Ryujo* and *Junyo*.

4 June42. **Battle of Midway**. Concentrating on the destruction of Midway air forces and diverted by U.S. horizontal, glide bombing, and torpedo attacks, the Japanese carriers are caught unprepared for the carrier air attack which began at 0930 with the heroic, but unsuccessful effort of Torpedo Squadron 8, and were hit in full force at 1030 when dive bombers hit and sank the carriers **Akagi, Kaga,** and **Soryu.** In the late afternoon, U.S. carrier air hit the Mobile Force again, sinking **Hiryu,** the fourth and last of the Japanese carriers in this action.

5June42. Planes from the Japanese carriers *Ryujo* and *Junyo* repeat their attack on installations at Dutch Harbor, Alaska.

3July42. USAAF B-24s bomb and damage the Japanese seaplane carriers *Kamikawa Maru* and *Kimikawa Maru* off Agattu, Aleutian Islands.

24Aug42. **Battle of the Eastern Solomons**. A Japanese attempt to reinforce Guadalcanal with a force of 58 ships, including three carriers and eight battleships. Planes from *Saratoga* sink the Japanese light carrier **Ryujo,** damage seaplane carrier *Chitose*, and destroy 90 enemy planes causing that force to withdraw.

1 Sept42. USAAF B-17s bomb and damage Japanese flying boat support ship *Akitsushima*

24Sep42. USAAF B-17 damages Japanese seaplane carrier *Sanuki Maru* off Shortland Island, Solomons.

Submarine *Trout* (SS-202) torpedoes Japanese escort carrier *Taiyo* east of Truk.

28Sep42. Submarine *Sculpin* (SS-191) torpedoes Japanese seaplane carrier *Nisshin* east of Kokoda Island

11Oct42. **Battle of Cape Esperance.** Japanese transport force, formed around seaplane carriers *Chitose* and *Nisshin* and six destroyers, reaches Tassafaronga, Guadalcanal, to disembark elements of the Japanese Army's 2nd Infantry Division.

15Oct42. Off San Cristobal, Solomons, planes from Japanese carrier *Zuikaku* sink the destroyer ***Meredith*** (DD-434).

26Oct42. **Battle of Santa Cruz Islands**. The fourth major carrier battle of 1942 is a costly win. The IJN is prevented from supporting a Japanese Army attack on Henderson Field, Guadalcanal. SBDs from *Enterprise* (CV-6) damage carrier *Zuiho*. SBDs from *Hornet* (CV-8) damage carrier *Shokaku*. *Enterprise* is damaged by planes from carriers *Junyo* and *Shokaku*. *Hornet* is damaged by planes from *Junyo*, *Shokaku*, and *Zuikaku*. *South Dakota* (BB-57) and *San Juan* (CL-54) are damaged by planes from *Junyo*. ***Hornet*** is abandoned.

Seaplane Carrier
- A ship type not in the USN inventory.
- 14 ships of 6,800 to 19,000 tons.
- Primary use to support troop landings. Also scouts.
- Some were dual purpose, oiler and seaplane carrier.

CHITOSE

IJN Chitose 12,000 tons, 630 feet, 25 seaplanes, 4 catapults, five hoists, 4- 5"guns, 29 knots.

The First Days of Guadalcanal 7 Aug - 20 Aug 1942

Operation Watchtower, the Tulagi, Guadalcanal, Santa Cruz invasion, was rushed to stop completion of a Japanese airfield that would interfere with supply lines to Australia. Marine planning was to attack in the Solomons in 1943. Instead, the operation had only five weeks to prepare for an assault on 7Aug42 with inadequate material and practice.

One practice amphibious landing was made at Fiji 28July for about 2/3 of the 1st Marine division to familiarize the troops with boarding landing craft and in unloading cargo. All accounts say the test was a failure. Also, transports and troops direct from the U.S. did not practice and were not packed for combat unloading, there was not time to reload every transport. The Marines departed Fiji on 31 July. The transit of 82 ships of the invasion force were not sighted because of bad flying weather. Landings were made in five areas from 475, 36-foot Higgins boats at 8 a.m. Friday, 07Aug42, D-Day.

The actual landing was easier than expected. Guadalcanal went virtually unopposed -- the 600 man construction crew ran to the interior when confronted with a thousand U.S. Marines. They did not see that another 10,000 Marines also come ashore ; the Japanese high command thought they were facing only 1,000 men on Guadalcanal for some weeks.

The existing Japanese seaplane base across the sound on Tulagi and the islands of Gavutu and Tanambogo were vigorously defended. Tulagi was taken Saturday afternoon, D+1, and the other two islands by late that night overcome by 6,000 additional Marines. The 1,500 defenders were killed with 108 Marine dead.

Red Beach at 1500 on 8 August (D +1), hopelessly blocke

However, all was not well. Unloading was continually delayed. Land-based bombers and torpedo planes attacked in mass within hours. Unloading of landing boats was assigned to 490 men of 1st Pioneer Battalion – exhausting work lifting cargo over gunwales -- only 115 Higgins boats had bow ramps. The few amphibian tractors proved their worth by being able to go directly from ship-side to inland supply dumps.

Hundreds of idle Marine reserves were not released to help unload or even to move material from Red Beach to inland dumps. Instead, ships had to send shore parties to help at the expense of ship antiaircraft defense.

Enemy air attacks began within hours. Twenty-five twin-engine Betty bombers with 19 Zero fighter escort from Rabaul attacked the ships by the early afternoon. Warned by coastwatchers, unloading stopped and the ships got underway to maneuver and fight off the attacks. A second attack occurred an hour later -- three hours of unloading time was lost.

One destroyer was damaged, *Mugford* (DD-389). Twenty-one U.S. Navy fighters were lost in the defense.

Morning of 8 Aug, D+1, saw an attack by 40 more bombers. **George F. Elliot** (AP-13) was set afire and *Jarvis* (DD-393) damaged. Both were lost. Twice more ships had to get underway and more unloading time was lost.

The landing beach became congested as the men tired. By nightfall, 150 landing boats were beached or offshore unable to unload ; ship discharge stopped.

With the landing a success and because of the large number of enemy land-based bombers and torpedo planes attacking the area and the fact that carriers were vital to maintaining the supply line from U.S. to Australia, the U.S. carriers were withdrawn with orders to move north of latitude 10°S only in pursuit of attacking fleets or of convoys.

Turner called Vandegrift (Marines) and Crutchley (cruisers) to a midnight meeting to discuss taking the rest of the vulnerable cargo ships out of harm's way the next day. The original schedule was to withdraw on D+4. Unloading delays and early departure meant that some cargo would go unloaded. 17,000 Marines had landed with provisions for one month and ammunition for four days of intense fighting. Some headquarters, artillery, and radar units did not get landed and almost none of the heavy construction equipment. 2,000 men in floating reserve, intended to take Ndeni on D+3, also withdrew. That island had been occupied by a seaplane tender, *McFarland* (AVD-14).

Just after midnight, a Japanese cruiser column arrived intending to attack the still unloaded transports. They ran into an Allied screening force of heavy cruisers near Savo Island. Inexplicably, the screening force of eleven ships, 5 CA, 6 DD, was surprised by the eight attackers, 5 CA, 2 CL, 1 DD. Four Allied heavy cruisers were sunk in about a half hour. Lost were **Astoria, Vincennes, Quincy, and HMAS Canberra**. The Japanese had expended much of their ordinance and with their schedule delayed, called it a night and withdrew without attacking the now thinly defended transports in the Sound, protected only by heavy cruiser HMAS *Australia*.

The three U.S. carriers withdrew before dawn Sunday as they were no longer needed for close air support of the successful landing and unaware that the cruiser screen had been attacked and destroyed. The transports withdrew that afternoon.

Operational planning had the primary objective to take an offensive airfield away from the Japanese and only secondarily to create an airfield for the Allies. Heavy equipment, aviation fuel, and specialized equipment were provided, but as low priority and that which was included did not get unloaded before the cargo ships departed. 400 drums of aviation gas were there, but these were probably intended to service the scout planes aboard the cruisers that were destroyed at Savo. The need for land-based air cover was recognized with the withdrawal of the carriers.

Marines remember this departure to this day as the "great Navy bug out", even though no significant ground fighting took place around Guadalcanal until 21Aug after 916 Japanese troops landed from 6 DDs on 18Aug. (Small unit activity took place from 12Aug.)

The Navy's responsibility was to spend the small U.S. fleet as expensively on the enemy as possible. This did not include having the carriers sitting around while twin-engine torpedo bombers tried to sink them. Fletcher moved off to be prepared for the inevitable Japanese attempt to push the Marines off the island, which was attempted two weeks later with a fleet larger than Fletcher's. He vigorously attacked a superior force in the Battle of the Eastern Solomons and sank his sixth enemy carrier, forced the enemy troopships to turn back, and saved the Marines.

A convoy of four destroyer-transports landed aviation fuel, bombs, and ground crews on 14Aug : *Calhoun, Gregory, Little,* and *McKean*. A second convoy with 3 APDs brought 120 tons of rations 20Aug. One converted merchantman sank on the way (it was top heavy from a concrete shield and turned over). Three of the four APDs were sunk within two weeks: **Calhoun, Little**, and **Gregory**. And *MV Lakatoi*, also top heavy, sank 21Aug in heavy weather.

Henderson field was declared operational on 17Aug (a PBY had landed briefly 12 Aug to test the field.) That same day there was a diversionary raid on Makin Island made by Carlson's Marine Raiders.

Naval aircraft arrived 20Aug ferried by *Long Island* (AVG-1) : 19 F4F Wildcats fighters and 12 SBD Dauntless dive-bombers. Air resupply and evacuation with R4D (C-47, DC 3) transports also began that day.

The first land battle on Guadalcanal, the Battle of the Tenaru River, began that night, at 01:30, 21Aug, in which almost 1,000 enemy were killed.

The Marine's first battle was a victory and they followed it up with victory after victory in coming months, but not a surprise. Other Marines had also fought heroically in the unsuccessful defense of Wake Island and the Philippines.

Both sides rushed to reinforce their part of the island in the coming months; the U.S. managed to stay a step ahead. Great naval battles that determined the course of the war were to begin in two weeks as the Japanese Combined Fleet, bringing new troops, counterattacked in the Battle of Eastern Solomons and the Battle of Santa Cruz. Skylark Channel became Iron Bottom Sound from all the sunken ships during October and November. Guadalcanal was secured 7 Feb 1943 after six months of battle with 24,000 Japanese and 1,600 U.S. ground and air forces dead.

Guadalcanal is remembered by Marines and in the public mind as a jungle battle, their first victory, and all the stories of personal and unit heroism wrap around those interests. Actually, it was a sea battle in which 48 warships went down, plus auxiliaries, half ours, half enemy : 3 carriers, 2 battleships, 12 cruisers, 25 destroyers, 6 submarines. USN had over 3,200 men killed. Of these, 1,270 Allied sailors died in the very first of six great sea battles there, more than the total Marine count in six months of fighting. The total lives lost at sea remains unknown ; the Japanese did not keep records of sailors killed or of the thousands of soldiers lost at sea when their transports were sunk.

The Allies were outnumbered by 10 carriers to 4, by 12 battleships to 8, yet the Navy kept reinforcements from overwhelming the Marines. Postwar interviews with surviving Japanese high command date their loss of the Pacific War to the inability to retake Guadalcanal. U.S. commanders agreed.

Grumman F4F Wildcat carrier f ighter

Japanese Betty Bombers Attacked Within Hours

Troopship *George F. Elliot* (AP-13) was set afire and destroyer *Jarvis* (DD-393) can be seen burning in the background.

G4M "Betty" bomber

Forty of these armed with torpedoes and a 3,000 mile range came looking for Fletcher's carriers the morning he withdrew just in time.

The Battle of Savo Island - Aug 9, 1942
Off Guadalcanal, Solomon Islands

In the summer of 1942, the Japanese had to be stopped in their drive to cut off Australia by severing the U.S. shipping lanes.

To this point in the Pacific War, the Japanese had destroyed the U.S. battle fleet at Pearl Harbor; destroyed the U.S. Asiatic Fleet in the Philippines; sunk the combined American British, Dutch, and Australian fleet in the East Indies (Java); punished the British fleet in Malaya and Ceylon and pushed the Indian Ocean fleet back to Africa; captured southeast Asia, the Philippines, the resource rich East Indies, and many island chains for defense in the central Pacific, an outpost in the Aleutians in the North Pacific, and Rabaul in the Bismarcks in the South Pacific. The advance to the south on Australia by way of New Guinea had been stopped by Admiral Fletcher at the Battle of the Coral Sea and the eastern Pacific was saved at the Battle of Midway. The Imperial Japanese Navy, even after the losses at Midway, still outnumbered the naval forces of the combined U.S. Pacific and the Australian fleets. The Japanese continued to progress south to isolate Australia.

America had established a Germany first policy. Eighty-five percent of U.S. military production, shipping and supplies was devoted to the Atlantic Theater against Germany, Italy, and their allies, and to aid England and Russia. U.S. troops had started to arrive in the United Kingdom. The Pacific Theater was divided into North, Central, and South Pacific under command of the Navy (Nimitz) and the Southwest Pacific (Australia to Philippines) under the Army (MacArthur). These two areas shared the remaining 15% of war production along with the China, Burma, India Theater.

Nimitz had two major war aims in 1942

> Protect Hawaii and the West Coast of the U.S. with Midway Island as his first line of defense.
>
> Protect the shipping lanes to Australia.

The Australian sea lanes were a line from the West Coast to Hawaii and from Panama Canal to Samoa, Fiji, New Hebrides to Brisbane, Australia. The Japanese move down the Solomons would allow them to control the Java Sea and threaten America bases in New Hebrides and Australia itself. Fletcher had reacted immediately to the Japanese occupation of Tulagi where a seaplane reconnaissance base was established and had turned back the invasion force that was coming around to the south side of New Guinea that faced Australia. He then had to race north to the major air-sea battle at Midway. During this period, the Marine Corps had been building up forces in

New Caledonia south of the New Hebrides. When the Japanese started to build an airfield on Guadalcanal across Savo Sound from their base at Tulagi, the United States knew it had to act before the airfield was completed.

The Solomons are a double string of eight main islands and many small islands spread along 700 miles of ocean about 1,200 miles northeast of Australia. The island chain runs northwest to southeast with Bougainville in the northwest, New Georgia in the middle, and Guadalcanal in the southeast. Fighting for New Guinea is going on 700 miles to the west. Guadalcanal is 92 miles long and 33 miles wide and 700 miles southeast of Rabaul on New Britain. New Britain is part of the Bismarck Island chain which is a northwest extension of the Solomons. The waters between the Solomon Islands is called The Slot. Immediately north of Guadalcanal at a distance of about 20 miles is the 20-mile long Florida Island where the Japanese have established one of their several seaplane reconnaissance bases in the Solomon Islands at Tulagi. The eastern end of the 400-mile long Slot is Savo Sound named for tiny Savo Island. The entrance to Savo Sound from the east is Indispensable Straight leading to several narrow channels. The entrances from the west are the north and south passages around Savo Island.

On 7 Aug 1942, the United States committed to its first land-based counterattack. The Marines landed at both Tulagi and Guadalcanal, on both sides of Savo Sound. The established base at Tulagi involved heavy fighting but was captured in two days. The not yet completed installation at Guadalcanal was of mostly construction workers who ran; it was an easy landing.

The Japanese responded immediately with air attacks from their bomber bases in New Britain (Rabaul) from the north and fighter strips in the northern Solomons (Bougainville). U.S. carrier planes operating near the invasion fleet in Savo Sound provided defense. Thirty-three enemy were shot down in the first air raid with a loss of 12 U.S. planes, one U.S. destroyer crippled, and a transport, *George F. Elliot* (AP-13), set afire and beached. The IJN also sent the Eighth Fleet from Rabaul to attack the U.S. beachhead. This fleet, VAdm Mikawa, consisted of five heavy cruisers, two light cruisers, and a destroyer.

The western approaches to Savo Sound were guarded by a screening force of six heavy cruisers and six destroyers (the battle fleet had been destroyed at Pearl Harbor) in two groups covering both passages. Radar pickets were the destroyers *Blue* (DD-387) and *Ralph Talbot* (DD-390) deployed west of Savo Island. The south passage was defended by HMAS Australia (CA.84, flagship of RAdm Crutchley, RN), HMAS *Canberra* (CA.33), USS *Chicago* (CA-29), *Bagley* (DD-386) and *Patterson* (DD-392). The northern group was made up of *Vincennes* (CA-44), *Quincy* (CA-39), *Astoria* (CA-34) and destroyers *Helm* (DD-391) and *Wilson* (DD-408).

The eastern approaches also had a screening force, made up of light cruisers *San Juan* (CLaa-54, flag), HMAS *Hobart* (CL.63), and destroyers *Monssen* (DD-436) and *Buchanan* (DD-484).

The IJN 8th fleet of fast cruisers arrived the second night and met the Allied screening force in the Battle of Savo Island. At the same time, the three U.S. carriers and their escorts, including *North Carolina* (BB-55), six cruisers, and 16 destroyers, were preparing to withdraw to get out of sight of the land-based, torpedo carrying bombers from Rabaul.

The enemy force of fast cruisers sent out scout float-planes that reported the American forces. Both radar picket ships (radar range about 10 miles) were at the extreme ends of their patrols sailing away from the Japanese fleet which passed undetected about 500 yards from *Blue*. The enemy was lost in the visual and radar shadow of nearby Savo Island. Allied ships were faintly silhouetted by a transport burning far over the horizon. The enemy discovered the southern force and fired torpedoes before they were detected.

Simultaneously with the explosions, the scout plane dropped flares illuminating the Allied fleet. *Canberra* was struck by two torpedoes and heavy shelling. The U.S. ships fired star shells and opened fire. *Chicago* of the southern force was torpedoed. The Japanese force turned north in two columns. The northern defense force had not gotten the word, there was a rain squall in the area, and they assumed the southern force was shooting at aircraft. The two Japanese columns passed on each side of the U.S. force and opened fire on *Astoria*, *Quincy*, and *Vincennes*. There had been no warning of enemy surface forces in the area – the American captains ordered "cease fire" assuming they were friendlies firing on their own ships. *Vincennes* caught a torpedo. *Robert Talbot* came charging south and was attacked first by friendly fire and then raked by the enemy escaping to the north. **Quincy** and **Vincennes** went down. During rescue operations for *Canberra*, *Patterson* was fired on by *Chicago*. **Canberra** was sunk the next morning to prevent capture as the U.S fleet left the waters that were hereafter called Iron Bottom Sound. *Astoria* sank about noon while under tow. *Chicago* had to undergo repair until Jan'43.

In just 32 minutes the enemy had inflicted massive damage. Four heavy cruisers were sunk and a heavy cruiser and destroyer badly damaged. 1,270 men were killed and 708 injured. The enemy had comparative scratches on three cruisers.

The rush to attack Guadalcanal can in no way let the Navy off the hook for the disaster at Savo Island. The fleet had been at war for eight months, including the Coral Sea and Midway, yet was walloped six to nothing.

What Went Wrong?

A court of inquiry determined that U.S. ships required more training in night fighting. *Dah!*

There were several sighting of the IJN 8th fleet by USAAF and RAAF aircraft along with several other Japanese ship movements : each report was of different ship compositions and bearings. Weather and enemy air defenses were a factor, yet a common denominator of these sightings was delay in getting the information from MacArthur's Army zone to Nimitz's Navy zone on the scene. Japanese seaplane carriers were included in the sightings and the Allied fleet prepared for a submarine or air attack, rather than surface action. Almost two thousand men paid for a chain of errors.

The Japanese 7th fleet cruiser's floatplanes were noticed and reported. Radio was poor that night and nobody associated aircraft reconnaissance with a surface attack. Visibility was 2 to 6 miles with rain in the area.

Both radar picket ships on patrol happened to be sailing away from the Japanese fleet. *San Juan* had modern search radar but was at the other end of the Sound. Was too much or too little reliance placed on this new technology? This battle must be considered to have been fought in the pre-radar days.

RAdm Crutchley, RN, was in command of the combined screening force in recognition of Allied unity : three of the eight cruisers were Australian. He had fought with Fletcher at the Coral Sea, but was not totally integrated with the U.S. Navy. *HMAS Canberra*, for instance, did not have TBS (short range radio known as Talk Between Ships) and could not hear the initial alarm issued by *USS Patterson*. Crutchley had left with his flagship, heavy cruiser *Australia*, that night to attend a conference called by RAdm Turner and did not participate in the battle. *Chicago* had the senior captain, but his ship was immediately torpedoed into a state of confusion that even included an exchange of friendly fire.

What went right?   Well, nothing, but luck helped a little.

Fortunately, the Japanese did not steam through and attack the now thinly defended transports. When the lead flagship turned towards the channel, his column, intent on sinking cruisers, failed to follow and continued north, then west to avoid shoal water, but away from the transports. The flagship then turned to chase after his squadron. To reform the Japanese fleet would have taken two hours; after attacking the transports and defenders the Japanese fleet would still be in the channel as daylight exposed them to carrier aircraft and any surviving ships of the earlier battle. The flag chart room had been destroyed so that navigation into the channel would have been dangerous.

Japanese naval tradition called for attacking warships. To expose cruisers to a second attack, with no torpedoes left, to extreme risk, for half-empty transports may not have seemed worthy. They had already won a great victory over warships and that was enough for one night's work.

The heavy cruiser, HMAS *Australia*, with screen commander Crutchley aboard, returning from his midnight meeting with Turner, was steaming to the battle site. Close support for the transports consisted of antiaircraft light cruiser *San Juan* and light cruiser HMAS *Hobart* and destroyers *Monssen* and *Buchanan*.

Unaware of the nature of the battle, VAdm Fletcher's force of 3 carriers and one battleship was returning from nighttime retirement and was not in range to attack the withdrawing enemy cruisers at first light. Fortunately, the Japanese did not know this. Equally fortunate was that an enemy air attack of 40 bombers carrying torpedoes early the next morning could not find the carriers and were only able to finish off *Jarvis* (DD-799) which had been torpedoed during the previous day's noon air attack. This day's search for carriers keep bombers from hitting the unprotected supplies stored just inside the treeline.

Afterward

All agreed the Japanese had not lost their fighting spirit after their defeat at Midway and that the allies had lost a major fight from problems with reconnaissance, communication, and preparedness. Yet RAdm Crutchley calls our attention that the propose of the screening force was to protect the landing and that the enemy did not get through. The cost was 1,270 sailors killed in one day, *more* than the Marine loss in the entire 6 month Guadalcanal campaign of 1,207.

F1M "Pete" Japanese Float Plane
launched from cruisers and battleships

Exploding a Myth about Tea
The Hudson Reported, but
the message did not get through in time.

We have had the privilege of spreading the truth and correcting a slur upon our Australian allies. These are notes associated with correcting this point in history.

I had repeated that an Australian pilot of a Hudson reconnaissance plane did not report his sighting of Mikawa's cruiser force approaching Guadalcanal, but rather, had tea on his return before debriefing. I was told by an Australian officer who was there that, "This is perpetuating the myth started by Morison in his 15 volume *History of U.S. Naval Operations.* Over the course of years I have corresponded directly with the crew and established that :

1. They did immediately break radio silence and report the sighting of the approaching Japanese fleet,

2. They did immediately return to base and make a full report when there was no confirmation of their sighting report; and

3. The U.S. admirals did not get the warning until the attack was underway. Following points 8 and 10 best explain what happened.

Notes:

1 . Operation Watchtower is the 1st Marine Division (MGen Vandegrift, USMC) landing on Florida, Tulagi, Gavutu, Tanambogo, and Guadalcanal in the first American land offensive of the war. Amphibious Force, South Pacific (RAdm Turner, TF 62), with screening force (RAdm Crutchley, RN) landed the leathernecks under cover of naval surface and air forces (VAdm Fletcher). Landings are supported by carrier aircraft (RAdm Noyes) and shore-based aircraft (RAdm McCain). Overall Commander South Pacific Force is VAdm Ghormley; Officer in Tactical Command is VAdm Fletcher.

2 . The Imperial Japanese Eighth Fleet was commanded by VAdm Mikawa with his flag on *Chokai* consisted of:
    Heavy Cruiser Division 6 (RAdm Goto) on *Aoba* with *Furutaka, Kako,* and *Kinugasa,*
    Light Cruiser Division 18 (RAdm Matsuyama) on *Tenryu* with *Yubari* and screen destroyer *Yunagi.*

3 . Americans understand "tea" to be an afternoon meal in which the beverage drunk is unimportant.

4 . The floatplane version of the Zero fighter is a Nakajima A6M2-N "Rufe" with a speed of 271 mph, range 1,100 miles, and armament of one 7.7mm and two 20mm; used in the Solomons.

5 . The Lockheed Hudson is a militarized Super Electra passenger liner (Army A-29 , Navy PBO-1) for maritime reconnaissance with a top speed of 246 mph, a range of 1,960 miles, and armament of 2 forward firing 7.7mm and usually a dorsal turret with two 7.7mm (.303 cal) and depth charges. The Solomons are 1,200 miles from Australia, so Fall River airfield was at Milne Bay, New Guinea, rather than near Townsville, Australia.

6 . The timezone changes in the vast Pacific is a worthy point. Also, the international dateline separates Washington and Hawaii from Tokyo and Sydney. Everybody reported in their own times, which confuse the record.

7 . The Pacific War zones of control were separated between Army (MacArthur in Australia) with the Southwest Pacific and the Navy (Nimitz in Hawaii) with the North, Central, and South Pacific. Operation Watchtower, a Navy operation, required moving the line of demarcation from 160 degrees west to 158 degrees west. The search area overlapped command responsibilities in addition to time zones.

8 . In defense of Morison (Naval Historian): two Hudsons saw the task force and the information did not get to the fleet off Savo Island. Delete the part about going to "tea" and replace it with **the radio station was off the air from an air raid alert** and an ignorant debriefing officer. Further study might even find a reason for the apparent arrogance of the debriefing officer; few servicemen in combat zones are intentionally obstructive. Note: the Aussie fliers had not been told that the American invasion was going on.

9 . LCdr Gregory's web page makes note that USAAF B-17s also saw the Japanese 8th fleet. McCain had responsibility for both U.S. Army and Navy shore-based aircraft in the South Pacific, but the Australian planes belonged to SW Pacific area, a separate chain of command.

10. Australian planes of MacArthur's SW Pacific command were flying out of an emergency field on the tip of New Guinea, the Hudson's Sighting Report path was -- **by radio to Milne Bay, from there to Port Moresby, to Townsville, to Brisbane ; then by motorcycle to MacArthur's HQ, by telephone to Canberra, and finally by radio to the fleet at Pearl Harbor for broadcast (on Fox) to all ships ; decoding, and delivery to the flag officer along with dozens of other messages. The elapsed time until Turner received the message was about seven and a half hours**, during which time the enemy advanced and Fletcher withdrew the carrier fleet from the confines of the Solomons into the Coral Sea.

Lockheed Hudson reconnaissance bomber

The Hudson was a passenger liner design, Lockheed Super Electra, converted for maritime surveillance. Many were sold to Commonwealth countries before the U.S. entered the war and helped bring American industry out of the depression.

    Max speed 246 mph, Range 1,960 miles, Bomb load 1,400 pounds (about 3 depth charges). A total of 2,584 Hudsons were built.

Nakajima A6M2-N "Rufe"

A "Zero" with pontoons, this plane was used in the defense of seaplane bases in the Aleutian and Solomon campaigns. Also used by some seaplane carriers and merchant raiders.

    Speed 270 mph, Range 1,000 miles, 327 were built.

IJN Heavy Cruiser *Chokai*, Flagship at Savo Island.

Chokai participated in many more night engagements in the Solomons. In the Battle off Samar, the Philippines, she attacked the U.S. escort fleet,"Taffy 3", and was stopped by a torpedo from heroic *Roberts* (DE-413), then bombed and scuttled. All survivors were lost when her rescue ship was then sunk.

Photo # 19-N-39212 USS Chicago off the Mare Island Navy Yard after her last overhaul. 20 December 1942

USN Heavy Cruiser *Chicago* had senior Captain at Savo Island

Chicago (CA-29) was assigned to the Australian Squadron, or MacArthur's Navy, where she participated in turning back the invasion of Port Moresby. She was part of the Tulagi, Guadalcanal operation defending at the disastrous Savo Island. Later, after repairs, *Chicago* was hit by two air-launched torpedoes while escorting Army troops, Jan 29, 1943. The next day while under tow, she was hit by four more, ending her career.

U.S. Naval Aircraft 1942

Fighters

The naval air service had just graduated from biplane aircraft in the 1936 fighter contract competition. That second wing had allowed the extra lift traditionally needed for carrier take off and Grumman F3F bi-planes were delivered through May 1938. An upgraded design by Grumman, an F4F, another bi-wing, was rejected before the competition and it was hurriedly redesigned with a wing removed as the F4F-2. The 1936 completion was between the Brewster F2A and the Grumman F4F-2. The F2A "Buffalo" won the single wing competition and units were introduced to front line service starting June 1939. War had started in Europe; much of the production order was sent to Finland. A second batch went to the RAF and to the Netherlands East Indies.

Brewster F2A Buffalo

A renewed completion awarded a contract for an improved single wing version of the Grumman F4F-3 "Wildcat." That model was being introduced to fleet service at the time of Pearl Harbor. F2A Buffalo's were in service at Pearl Harbor, in the Philippines, Wake Island, Ceylon, and Midway -- where no Allied planes were a match for the Japanese Zero fighter. There were 140 Grumman F2F and F3F bi-planes serving as fighter pilot trainers in December of 1941. Grumman continued development with the F4F-4 where production was provided by General Motors under the designation FM-1. The F4F-8 was produced by General Motors as the FM-2. Meanwhile, Grumman concentrated on the next generation fighter, the F6F Hellcat, that entered service in late 1942 and qualified pilots entered combat in Aug 1943 with a fighter of superior performance to the Zero.

The Japanese introduced a new fighter in China in 1940. Reports as to its specifications were not believed in Washington, preferring to believe that it was Chinese incompetence, rather than a superior Japanese fighter that was causing Chinese losses. 78 Zeros flew in the attack on Pearl Harbor.

Designation	F2A Buffalo	**F4F-3 Wildcat**	A5M Claude	**A6M2 Zero, Zeke**
Manufacturer	Brewster	Grumman	Mitsubishi	Mitsubishi
Contracted	1934	1938	1935	1940
Flew	Dec 1937	Mar 1939	Feb 1935	Apr 1939
Service	1939-1942	1940-1943	1936- 1942	1940-1945
Engine - hp	1200	1200	710	925
Max mph	320	318	280	346
Guns - caliber	4-.50	4-.50	2-.30	2-.30, 2-20mm
Range - miles	965	770	746	1,120
Weight - lbs	4,723	5,760	2,680	4,178
Number Produced	161 + 354 export	Mod-3, 560 -4, 1,320 FM-2, 4,777	1,394	11,000

Grumman F4F-3 Wildcat

Grumman F4F-4 Wildcat

Torpedo Bomber.

American torpedoes were defective in 1942, while other nations had great success with torpedo attacks in the Mediterranean and, of course, at Pearl Harbor. The Douglas TBD Devastator was the first U.S. single wing, metal, enclosed cockpit torpedo bomber. It was contracted in 1934 and introduced in 1937. 130 were built and it was the fleet torpedo bomber until mid-1942. The TBD served in the attacks on Lea and Tulagi and at the Battle of the Coral Sea. At Midway, a month later, the slow TBDs were decimated in uncoordinated attacks, 35 out of 41 deployed did not return. The Navy awarded contracts for competition for a new torpedo bomber in 1940 to Grumman, TBF, and Vought, TBY. Each flew in Aug 1941. The Grumman design won and went on to production with such demand that General Motors provided production, as the TBM, while Grumman concentrated on fighter production. The new TBF "Avenger" was just reaching the fleet by mid-1942 and the Battle of Midway, but pilots had not yet been carrier qualified in it. Six flew from the Midway Island field with 5 of the 6 shot down by Zeros before reaching their targets.

Forty Japanese Nakajima B5N "Kate"s were used at Pearl Harbor as torpedo bombers ; 103 were used as horizontal bombers. During 1942, Kates also stopped *Lexington, Yorktown,* and *Hornet.*

Designation	**TBD**	**TBF/TBM**	**B5N**
Manufacturer	Douglas	Grumman/ General Motors	Nakajima
Name	Devastator	Avenger	Kate
Contracted	1934	1940	1935
Flew	Apr 1935	Aug 1941	Jan 1937
Service	1937-1942	1942-1946	1937-1944
Engine	900 hp	1,700 hp	1,000 hp
Max Speed	206	269	235
Cruise speed	126	146	140
Range, miles	435 w/ torpedo 716 w/ bombs	1,050	1,235
Weight	5,600	10,560	5,025
Number Produced	130	2,293 / 7,546	1,149

Dive Bomber.

The dive-bomber did prove effective, much as the Stuka did in the European theater, but only in mass attacks with significant air protection. The hit rate was low; 112 carrier sorties by Navy and Marine dive bombers at the Battle of Midway made only 13 bomb hits. A dive-bomber could maneuver with a ship and carry its bombs to the ship without requiring the bomb run and long free fall period of a high altitude horizontal bomber. The dive bomber, with its 2-man crew, was also the U.S. carrier-borne scouting aircraft.

129 "Val" bombers were in the attack on Pearl Harbor. "Val"s later sank HMS *Hermes* (CVL), *Cornwall* (CA), and *Dorsetshire* (CA) in the April 1942 raid near Ceylon.

Designation	SB2U Vindicator	SBD Dauntless	D3A Val
Manufacturer	Vought	Douglas	Aichi
Flew	Jan 1936	April 1939	1939
Service	1937-1942	1941-1945	1940-1944
Engine hp	825	1,350	1,000
Max Speed	243 mph	255 mph	267 mph
Cruise Speed	185 mph	183 mph	200 mph
Range	1,120	773 miles	913 miles
Weight	5,635	6,535 lb	5,309 lbs
Number Produced	170	4,000+	1,500
Crew	2	2	2
Guns	3- .50	2-.50, 2-.30	3- .30

Vought SB2U Vindicator

Douglas SBD Dauntless

Horizontal Bomber.
The U.S. had no specific, carrier-based horizontal bombers. Torpedo bombers could and did act in the role over land targets as well as to carry bombs and depth charges. The Japanese first wave at Pearl Harbor used 40 "Kates" as torpedo bombers and 49 "Kates" as horizontal bombers. The second wave had 54 "Kates", all as horizontal bombers. The Japanese Navy also had a shore-based air component, that had access to all of the carrier plane types as well as twin-engine bombers. The U.S. later incorporated multi-engine bombers as patrol planes with an attack capability.

Scout Float Plane.
Two to four float planes were carried by cruisers and battleships to search for enemy fleets and submarines, to provide gunnery spotting, rescue and other services. These were launched from catapults atop gun mounts or from amidships of WWI era ships and from the fantail of WWII build ships. On return, the seaplane was hoisted aboard by crane and repositioned on its catapult or hanger.

    Curtiss SOC Seagull in service from 1935 to 1946.   332 built
    Vought OS2U Kingfisher – Aug 1940 till helicopters. 1,500
    Curtiss SO3C SeaMew/Seagull – 1942-1944.   440

Reconnaissance Bomber - Flying Boats.
The Consolidated PBY Catalina was an all-purpose, air, sea, and land flying-boat; an amphibian version added wheels. As a long-range reconnaissance plane, it could range 2,545 miles and patrol for over 20 hours. It dropped bombs on shipping and depth charges on submarines. It carried out the only successful torpedo attack at Midway, damaging an oiler. PBY Catalinas, operating from the seaplane tender *Gillis* (AVD-12) in Nazan Bay, Atka Island, hit ships and enemy positions on Kiska in an intense 48-hour attack which exhausted the gasoline and bomb supply aboard the *Gillis,* but was not successful in driving the Japanese from the island. As effective as it was in reconnaissance and attack, it sometimes carried two lifeboats under its wings, rescue was an important part of the job.

Consolidated PBY Catalina

Reconnaissance Bomber - Land Based.

Long Range Bombers resulted in an Army-Navy controversy over coast protection. Traditionally the Navy provided coastal protection beyond the range of Army shore-based artillery. The Army Air Corps' BGen Billy Mitchell made dramatic displays that it could fly at long ranges, could find ships far at sea, and could destroy armored warships from the air. The Navy considered all coast defense that was beyond the range of artillery to be their domain. Air power aficionados considered the Navy outdated and motivated by self-preservation.

European air attacks early in the war showed that massed, land-based air power could destroy ships at sea. However, the American experience was far from convincing, primarily because there were not enough aircraft available to make massed attacks.

A single B-17 bomber could carry a lethal bomb load, but had to attack from high altitude to escape ship-borne antiaircraft fire. At high altitudes, a strategic bomber could not hit a maneuvering target. A bomber had to fly straight and level to the target to align the bombsight to where the bombardier thought the target would be after the bombs made a long fall. During that period, as the bomber was committed, a ship could maneuver for several minutes to confuse the airplane and to avoid the falling bombs. The Army B-17 was sold for coast defense before the war. The press was primed for B-17 success and airmen reported splashes as hits. In fact, in all of 1942, one Japanese destroyer was sunk, and it was stopped to pickup survivors from a ship sunk by carrier planes. B-17s did damage several warships and sank several transports in harbor or convoy. 1943 saw another hit on a destroyer, a seaplane carrier, and more transports.

Fletcher's time is before the period of American dominance with the:
- Chance Vought F4U Corsair - first combat with USMC on Guadalcanal 13Feb43 with 12,570 built.
- Grumman F6F Hellcat - first combat from *Yorktown II* 31Aug43 with 4,423 built.
- Curtiss SB2C Helldiver - first combat in the second strike on Rabaul 11Nov43 with 6,000 built.

The year 1943 saw Japan lose 6,203 planes (4,824 airmen), three times more than they had to start the war and, strangely, Japan did not have an adequate replacement training program.

Japanese Naval Aircraft

Designation	Dates	Speed / Range	Guns / Bomb Load	Built
Interceptors A5M "Claude"	1937-1942	270 mph 746 mile	2- 7.7 mm 2- 66 lb	1,100 fixed gear
A6M "Zeke" (Zero)	1941-1945	332 mph 1,118 mile	2- 20 mm; 2- 7.7 mm 2- 132 lb	10,450

Best performing fighter of any country in early WWII, but vulnerable. Continually upgraded, but the design was dated by mid-1943.

Japanese Army Fighters

Ki-27 "Nate"	1938-1943	290 mph 390 mile	2- .30	Similar to Claude, 3,400 built
Ki-61 "Tony"	1943-1945	348 mph 1,118 mile	2-.50, 2- 20mm; 2- 550 lb	Similar to Bf-109 2,750 built

Dive Bomber

D1A "Susie"	1934-1940	193 mph 580 mile	3- 7.7 mm 1- 550 lb or 2- 66	biplane, open, fixed gear
D3A "Val"	1940-1945	267 mph 900 miles	3- 7.7 mm 1- 550 lb, 2- 132	fixed landing gear 1,500
D4Y "Judy"	1943-1945	343 mph 2,417 mile	3- 7.7mm, 1- 1,100 lb, 2- 66	2,000

Torpedo Bombers

B2M "Ripon"	1932-1940	132 mph	2- 7.7mm; torpedo or 1,070 lb bomb	biplane, open cockpit
B4Y "Jean"	1934-1940	173 mph	1- 7.7mm; torpedo or 1,100 lb bomb	biplane, open cockpit
B5M "Mabel"	1937-1942	235 mph 1,400 mile	1- 7.7 mm; 1- 17.7 torpedo or 1,764 lbs	fixed landing gear 125
B5N "Kate"	1938-1944	235 mph 1,237 mile	1- 7.7 mm; 1- 17.7" torpedo or 1,400 lbs	1,150
B6N "Jill"	1943-1945	299 mph 1,890 mile	2- 7.7 mm; 1- 17.7" torpedo or 1,764 lbs	1,130

Notes on Japanese Aircraft Carriers

Carriers did not have catapults, therefore flight operations depended on speed into the wind as the only aid to launching. Some carriers converted from other designs were slow and had to use more flight deck for takeoff thereby interfering with both flight and storage operations.

The Japanese always suffered from gasoline fires when hit, whereas the American carriers learned to first empty gas lines, then to flush them with CO^2 or nitrogen to prevent secondary explosions.

During the lull in the naval air war, late'42 to mid'44, Japan should have built Shokaku class carriers, the equivalent of the Essex class, yet only one was build, *Taiho*. Shipbuilding resources were spent on the supercarrier *Shinano* and on conversion of ocean liners into escort carriers. In this period Japan launched 1 CV, 5 CVL, and 2 CVE. The U.S. launched 10 CV, 9 CVL, and 50 CVE.

Japan never considered anti-submarine activities for carriers. Meanwhile, USN submarines sank the carriers and merchant fleets that were the life of Japan. Japan started the war using their submarines successfully for anti-shipping and anti-fleet activities, but soon felt compelled to disarm much of their submarine fleet for use to supply remote islands. The USN paid much attention to anti-submarine defense, initiated by the Atlantic U-boat menace, and kept Japan's submarines suppressed. For example, 6 subs were sunk in 12 days by **ONE** U.S. destroyer escort, USS *England* (DE-635).

Japanese aircraft were designed from experiences of the war in China for attack against a weak foe. They had few defensive features. The Japanese Navy had both land-based and ship-borne wings.

Japan started the war with experienced pilots and a buildup of excellent aircraft. There were limited replacements for either. The U.S. introduced massive training programs and had high-tech industrial capacity. Japan's manufacture depended on many small machine shops and could not easily get new designs into production once they were geared to produce Zeros, thus they could not efficiently improve that plane or introduce improved designs. Changes throughout the war did slowly occur : increase engine from 925 hp to 1,130, shorten wing, improved 20mm guns, change from .30 to .50, increase speed from 316 to 351 mph.

A second generation Zero, the A7M "Sam" (390 mph) should have met the Hellcat in 1943, but only 8 prototypes were built before war-end. The Kawanishi N1K1 "George" (370 mph) with 1,450 built for service in 1944, but the carrier version, N1K4, never got beyond prototype.

The Mitsubishi J2M "Jack" (370 mph), also land-based, entered service in 1944, but only 475 aircraft could be built before the war was over.

American aircraft carried pilot armor and self-sealing gas tanks, whereas Japanese planes easily flamed.

Japanese pilots were willing to die and, early in the war, didn't even wear parachutes. Once the experienced pilots were gone, there were few well-trained replacements.

A6M "Zeke", Zero

D3A "Val"

B5N "Kate"

The Battle of the Eastern Solomon Sea
Aug 24. 1942
Saving Guadalcanal

The Japanese underestimated the number of Americans on Guadalcanal, but could not underestimate the need to remove them. Major elements of the Imperial Japanese Navy were sent to escort only 1,500 troops, which was thought to be sufficient to destroy the Marines on the island and any defending naval force as easily as they had destroyed the screening force at Savo Island.

The Imperial fleet comprised three carriers : both remaining fleet carriers, *Shokaku* and *Zuikaku,* and light carrier *Ryujo* ; three battleships, *Mutsu, Hiel,* and *Kirishima* ; 13 heavy cruisers ; 3 light cruisers ; seaplane carrier *Chitose* ; and 31 destroyers. These were escorting auxiliary cruiser *Kinryu Maru* with Naval Landing Force of 800 men and four APDs with an army detachment of 700 soldiers. Land-based air support was 100 operational aircraft ; underwater were 10 submarines.

The American defenders under VAdm Fletcher had three fleet carriers: [1] *Saratoga* (CV-3), *Enterprise* (CV-6), and *Wasp* (CV-7) ; one battleship, *North Carolina* (BB-52) ; five heavy cruisers, *Minneapolis* (CA-36), *New Orleans* (CA-32), *Portland* (CA-33), *San Francisco* (CA-38), and *Salt Lake City* (CA-25) ; two new antiaircraft light cruisers, *Atlanta* (CLaa-51) and *San Juan* (CLaa-54) ; and 18 destroyers.

Land air support was by 23 Marine planes that had arrived a few days before at Henderson Field; 39 PBY reconnaissance planes from seaplane tenders; and 30 B-17s from the New Hebrides. Three picket lines of submarines kept watch for the Japanese fleet.

Although the Japanese were known to be coming, they refueled at sea and arrived before projected by HQ in Pearl Harbor. *Wasp* was sent on refueling rotation; she and her task group escort of 3 cruisers and 7 destroyers ; one-third of the already outnumbered U.S. fleet, missed the fight.

The Japanese had the light carrier *Ryujo* ahead of the strike fleet. She opened the battle with an air strike on Henderson Field. Fletcher sent dive and torpedo bombers to attack and **Ryujo** was successfully sunk without loss. Meanwhile, scouts from both sides found each other's two big carriers. The ensuing battle was essentially a giant aerial dog fight interspersed with ship-borne antiaircraft fire. The U.S. lost 20 planes, the Japanese lost 70. *Enterprise* took a couple of bombs and *Chitose* was nearly sunk, but survived. There was superficial damage to *North Carolina* and *Shokaku*, but major damage to two Japanese light cruisers, a destroyer, and 2 submarines.

Till they met Fletcher

When steering control was returned to *Enterprise*, Fletcher retreated in the direction of *Wasp* for the night. The Japanese fleet, expert in night warfare charged forward and found nothing ; Japanese destroyers bombarded Guadalcanal.

Next morning Marine pilots from Guadalcanal attacked the transports and eight B-17s from Espiritu Santo sank a destroyer which had stopped to rescue troops. *Wasp* advanced to the scene but found nothing. Without air cover, the Japanese had retreated without delivering the troops or destroying an American fleet half its size. The third of the five great carrier battles of the Pacific War had ended with the Japanese minus a carrier, a destroyer, a submarine, and an air group.

Enterprise transferred her bombers to Henderson Field on Guadalcanal and returned to Tongatabu for temporary repairs before heading for Pearl Harbor.

The Japanese gave up on major fleet actions to deliver troops and initiated the *Tokyo Express* in which destroyers delivered reinforcements and supplies at night and retreated beyond U.S. air range by daylight.

---

(1) These are the carriers that Fletcher had saved the third day of the invasion of Guadalcanal that upset the Marines when the Navy sailed away two weeks before.

USS Wasp (CV-7) entering Hampton Raods, VA, May 1942

USS Enterprise (CV-6)

"The Big 'E'" was the top Navy warrior.

Participated in 20 battles in WWII and survived the war.

Milne Bay – Aug 26-Sept 5, 1942

At the same time the IJN failed to reinforce Guadalcanal, a Japanese naval landing force was put ashore on the tip of New Guinea to take Milne Bay. Initial advances were made during rain storms, but the Japanese invaders were outnumbered ten to one and with Australian aircraft able to attack when the weather cleared, the Japanese were forced to withdraw after the third day. The Japanese general staff decided that reinforcements must be sent to Guadalcanal. The Japanese military had overextended; the priority became to expel the Americans and no longer to take new land. That effort continued until February 1943 when they were forced to begin the long retreat from the Pacific. The New Guinea campaign resisted MacArthur until May 1944.

Aleutians – Sept 16, 1942

Up North, the Japanese transferred their garrison at Attu to Kiska, then returned two months later.

Proximity Fuse

The mini-radar proximity fuse was introduced to combat 5-months after the successful test in the Chesapeake Bay with a 7-fold increase in effectiveness. The imperial staff wondered at the amazing accuracy of U.S. gunners in the Solomons campaign. A special team of observers sent to examine the situation was itself shot down with proximity fused AA. The secret of the radar fused shell was so protected that they were not fired over land, so that a dud was not recoverable to reveal its secret.

Coast Watchers

The coastwatcher idea was initiated right after WWI where responsible men were recruited to report enemy activity. By 1942 most were in enemy territory. Those in the Solomons, mostly ex-planters, were the most notable. When the Solomons were occupied by the Japanese, the coastwatchers headed to remote observation posts with their radio transceiver and binoculars and lived off of the land

Coastwatchers Mason and Read, located at opposite ends of Bougainville, reported air and sea movements from Kavieng and Rabaul. The beachhead at Guadalcanal was given 50 minutes warning of aerial attack by number and types of aircraft. Carrier air and pilots from Henderson Field had time to gain altitude and vector to intercept. Also interesting was the count of fewer Japanese planes returning. The first attack warning on Guadalcanal was of 24 torpedo bombers which had only one return. The re-invasion convoy of 11 transports was reported in the Slot and only four ships made it to Guadalcanal and they had to beach themselves to keep from sinking. Adm Halsey credits the coastwatchers with saving Guadalcanal. The Japanese soon discovered the source of their problems and made life even rougher by forming local police forces to root out the coastwatchers.

Coastwatchers also assisted isolated detachments that had not surrendered. Many instances of rescue of downed airmen and sailors by "friendly natives" were actually auxiliaries sent by coastwatchers. The part played by the coast-watchers was suppressed for their protection. Natives were offered tinned meat or similar rewards for each hapless Allied member delivered live to the coast watcher ; 120 downed pilots were brought in. LTjg John F. Kennedy, *PT-109*, was rescued by Reg Evens, a coastwatcher who had spotted his boat on fire that night and effected a rescue when Navy ships and planes were unable to find the site. Yes, he was invited to the White House, in 1961.

When we read of the Navy's annoyance on being diverted to effect the submarine rescue of civilian, missionary and coastwatcher, we must think that these men that had put their lives on the line for months, whose location had been overrun, and needed a way out. The alternative resulted in capture of the Australian team and execution. They had paid their dues.

Submarine Attacks – Aug-Sept 1942

Imperial Japanese Navy submarines accompanied the Japanese fleet to the Solomons area. They did their damage after the surface fleet retreated from the Battle of Eastern Solomons, 24Aug42. Three were sunk : *I-123* on 28 Aug by *Gamble* (DM-15) ; *RO-33* on 29 Aug by HMAS *Arunta* (DD.30) ; and *RO-35* disappeared. The rest did significant damage to U.S. convoy escorts accompanying troops and supplies to Guadalcanal. The area was called "Torpedo Junction".

Saratoga Damaged -- 29 Aug 42

Continuing to patrol east of the Solomons, *Saratoga* (CV-3) was found by *I-26* that sent a spread of six torpedoes at *Saratoga*. One torpedo hit and damaged her turbo-electric propulsion system. *I-26* had been so intent on her attack, that her hull was grazed by *MacDonough* (DD-351) who had just detected her periscope and was on a depth charge run. *I-26* escaped. *Saratoga* returned to Tongatabu and then to Pearl Harbor for permanent repair until November. Fletcher, on *Saratoga*, had his first leave from combat since the war began.

Wasp Sunk and *North Carolina* Torpedoed -- 16 Sept 42

Wasp (CV-7), *Hornet* (CV-8), *North Carolina* (BB-55) with 10 other warships were escorting transports carrying the 7th Marine Regiment from Espiritu Santo to Guadalcanal as reinforcements. *Wasp* was performing flight operations with gasoline and munitions exposed. *I-19* fired a spread of six torpedoes at *Wasp* and hit her with three. Gasoline and ready ammunition ignited followed by internal explosions. She was ordered abandoned after a half hour and was sunk by her escorts that night.

About the same time, *I-15* was thought to have sent a torpedo spread on a line towards *Hornet* and *North Carolina* operating about 5 miles to the northeast. After the war records show she had not fired and these were the three *misses* on *Wasp* from *I-19* that had lined up the two task forces for her spread. These had missed *Hornet*, but one caught *North Carolina*. Two minutes later, *O'Brien* (DD-415) was shattered. *North Carolina* returned to Pearl Harbor for repair of a 20-foot hole and was out of action for the rest of the year. *O'Brien* was temporarily repaired, but *O'Brien's* damaged seams opened up a month later and she sank while returning to San Francisco for a permanent repair.

With *Enterprise* (CV-6) damaged by bombs at Eastern Solomons ; *Saratoga* damaged by a torpedo ; and *Wasp* sunk ; *Hornet* was the only carrier left in the South Pacific for six weeks. Then *Hornet*, too, was lost in the Battle of Santa Cruz Island on 26 Oct 1942 from air attack. *Enterprise* was damaged, again, and there were **no active fleet carriers in the Pacific** until *Enterprise*

Till they met Fletcher 157

returned 12 Nov for the Naval Battles of Guadalcanal with repair parties still aboard and one elevator out of service.

Japanese Use of Submarines.

The *Tokyo Express* of destroyers ran troops and supplies to Guadalcanal by night. Eventually, they could not support the 15,000 men they had delivered to Guadalcanal. Submarines, too, were assigned to the task. By year-end, thirty-eight combat submarines were stripped of arms and used for resupply. Twenty of these were sunk in the effort. Four fleet attack submarines were also sunk during the period. The few offensive submarines in the area, while damaging enough, allowed the rapid buildup of U.S. troops and material.

Meanwhile, the USN worked on its problems with defective torpedoes and offensively attacked every enemy ship sighted.

By the end of the war, Japanese submarines could not keep up with supplying outposts on remote islands that had been bypassed by U.S. forces Survivors were only saved from starvation by the end of the war when U.S. ships raced there to accept their surrender and also to provide food.

USS Saratoga (CV-3) in good times and bad.

Battle of Cape Esperance, night of Oct 11-12, 1942
Cruiser Victory, sorta

A newly formed cruiser force, Task Group 64.2 (RAdm Norman Scott), consisting of two heavy cruisers, *San Francisco* (CA-38) and *Salt Lake City* (CA-25) ; two light cruisers, *Helena* (CL-50) and *Boise* (CL-47) ; and five destroyers were protecting the southwest approaches in the landing of Army troops at Guadalcanal. Other Task Groups were based around *Hornet* (CV-8) to the west and around *Washington* (BB-56) to the east.

Scout aircraft sighted a "Tokyo Express" and the task force sailed around the western end of Guadalcanal to block the entrance to Savo Sound. When the enemy force of three heavy cruisers and two destroyers was sighted by radar, Scott reversed course to cross the enemy's "T" leaving the van destroyers in the rear racing to catch up. They were passing between the Japanese and American task forces when both fleets opened fire. The destroyers made an unsuccessful torpedo run and the American cruisers temporally held fire to identify the destroyers in the dark. The Japanese reversed course exposing each ship to fire as it turned. The Japanese heavy cruiser, **Furutaka,** and destroyer, **Fubuki,** were sunk and another CA and DD were damaged. USS *Duncan* (DD-485) was disabled by enemy fire and further damaged by friendly fire and was abandoned just before she blew up. *Salt Lake City*, *Boise,* and *Farenholt* (DD-491) were damaged.

Meanwhile that night, a Japanese transport force formed around seaplane carriers *Chitose* and *Nisshin* and six destroyers, reached Tassafaronga, Guadalcanal, to disembark elements of the Japanese Army's 2nd Infantry Division. Other U.S. Task Groups do not seem to have been engaged.

USS Duncan (DD-485)

Battle of Santa Cruz - 26 Oct 1942
Final Carrier Battle of the South Pacific

The Battle of Santa Cruz was to be the last of the carrier battles around Guadalcanal. Both sides had to withdraw their mauled carriers. This battle set up conditions for the surface fleet actions in the Naval Battles of Guadalcanal in mid-November.

The Japanese attack consisted of four carriers: *Shokaku* (CV), *Zuikaku* (CV), *Junyo* (CV), and *Zuiho* (CVL); four battleships: *Hiei, Kirishima, Kongo*, and *Haruna*; nine cruisers; 28 destroyers; eleven submarines and seven other ships.

On the U.S. side was a fleet of less than half that size: *Enterprise* (CV-6), *Hornet* (CV-8), *South Dakota* (BB-57), *Portland* (CA-33), *Northampton* (CA-26), *Pensacola* (CA-24), *Juneau* (CLaa-52), *San Diego* (CLaa-53), *San Juan* (CLaa-54), and 14 destroyers. The light cruisers were new, very light, antiaircraft types with sixteen 5" guns, rather than the fifteen 6" guns carried by most light cruisers.

Enterprise had departed Pearl Harbor, Oct 16, for the South Pacific, repaired from the big air battle at Eastern Solomon Sea on 24Aug. *Hornet* had borne the brunt of air cover in the Solomons until 24 October 1942 when she was joined by *Enterprise* northwest of the New Hebrides Islands. Together the two carrier groups formed TF 61 (RAdm Kinkaid) and steamed to intercept a Japanese carrier and battleship force bearing down on Guadalcanal to support a Japanese Army assault on Henderson Field. Scout planes located the enemy fleets and each attacked. The Battle of Santa Cruz Island took place 26 October 42 without contact between surface ships of the opposing forces.

That morning *Enterprise* planes bombed carrier *Zuiho.* Planes from *Hornet* severely damaged carrier *Shokaku,* and cruiser *Chikuma.*

Meanwhile, **Hornet** was fighting off a coordinated dive bombing and torpedo plane attack which left her so severely damaged that she had to be abandoned. Destroyers *Mustin* and *Anderson* attempted unsuccessfully to sink the burning hulk with nine torpedoes and shell fire. Japanese destroyers eventually sank her that night by firing four 24-inch torpedoes at her blazing hull.

Enterprise was hit twice by bombs with 44 killed and had 75 wounded. Despite serious damage, she continued in action and took on board a large number of planes from *Hornet* when that carrier had to be abandoned.

Porter (DD-356) stopped to pickup a downed aircrew and was torpedoed either by *I-21* or a ditching TBF. An *Enterprise* pilot dived to machine gun the torpedo but was not in time. **Porter** was abandoned and sunk by *Shaw* (DD-373) after that ship took off her crew.

That evening the American forces retired to the southeast. Although the battle had been costly, combined with the Marine victory on Guadalcanal, they had turned back the attempted Japanese parry in the Solomons. American losses of a carrier and a destroyer were more severe than the Japanese which lost none -- both sides had several badly damaged ships. The Japanese goal was not achieved and the battle gained time for the U.S. to reinforce Guadalcanal against the next enemy onslaught. Furthermore, the damage to two Japanese carriers and a major loss of aircrews sharply curtailed the air cover available to the enemy in the subsequent Naval Battle of Guadalcanal.

Enterprise entered Noumea, New Caledonia, for repairs by *Vestal* (AR-4). For several weeks there were *no* active U.S. carriers and only one battleship in the South Pacific. *Enterprise* sailed for the Naval Battle of Guadalcanal with repair crews still on board, launched her planes and retreated sending her planes to join the Cactus Air Force on Guadalcanal. *South Dakota* took a bomb on the forward gun mount but participated in the Naval Battle of Guadalcanal. *San Juan* took a bomb through the fantail; she repaired in Sydney and missed the next big battle.

Destroyer takes off crew from damaged ***Hornet*** **(CV-8)**

Hornet was new and on shakedown near DC when Pearl Harbor occurred. She was tapped to try launching Army B-25 "Mitchell" medium bombers and when successful sailed immediately for San Francisco, loaded 16 bombers, and sailed for Tokyo with Doolittle's Raiders aboard. On return she sailed to the Coral Sea, but arrived after that fight and was recalled to participate in Midway. With new radar installed she sailed south to become the only active carrier in the Pacific. At Santa Cruz she damaged a carrier and was herself so badly damaged that she had to be abandoned. Age, one year and six days.

Anti-Aircraft Light Cruisers

Four antiaircraft light cruisers were authorized in 1940 to mount the dual purpose 5"/38 gun. The Atlanta Class had eight double turrets, three stepped forward, three stepped aft and one on each side, midships. These were very light cruisers, 6,000 tons, whereas other WWII light cruisers were 10,000 tons and heavy cruisers were 13,600 tons. The Atlanta Class turned out to be top heavy and the next two classes of antiaircraft cruisers did not have the two mid-ship turrets. The third batch, delivered post-war, also had the four middle turrets lowered one deck to further reduce the center of buoyancy.

The needs of war required that *Atlanta* (CLaa-51) and *Juneau* (CLaa-52) participate in the night action of First Naval Battle of Guadalcanal in which *Atlanta* sank the next day from damage and the already damaged *Juneau* was torpedoed during the withdrawal and went down in seconds with the Sullivan brothers. *San Diego* (CLaa-53) provided protection to *Enterprise* (CV-6) during that three-day battle. *San Diego* went on earn 15 battle stars and was honored to be the first USN ship to enter Tokyo Bay on 27Aug'45. *San Juan* (CLaa-54) was escorting *Saratoga* (CV-3) to the South Pacific during that Naval Battle of Guadalcanal. Damaged twice early in the war, *San Juan* finished the war with 13 battle stars and also entered Tokyo Bay 27Aug'45.

The second batch of antiaircraft cruisers, the Oakland Class, arrived about two years later with six dual 5"/38 turrets and many 40mm and 20mm : *Oakland* (CL-95), *Reno* (CL-96), *Flint* (CL-97), and *Tucson* (CL-98). They proved handy with Kamikazes in the last months of the war.

The third batch, the Juneau Class, continued with six dual 5"/38 turrets, but with four of the turrets lowered one deck, 24- 40mm guns, and no torpedo tubes. These were commissioned in 1946 : *Juneau II* (CL-119), *Spokane II* (CL-120) and *Fresno* (CL-121).

USS *San Diego* (CLaa-53)

Naval Battle of Guadalcanal
12-16 November 1942
End of the Japanese hope for conquest.

The Japanese finally sent most of the remaining Imperial Fleet in a last attempt to expel the Americans from Guadalcanal.

Japanese Forces.
Available were 2 carriers, 4 battleships, 7 heavy cruisers, 4 light cruisers, and 30 destroyers. The carrier group (2 CV, 2 BB) provided distant cover.

USN Forces
The U.S. had *no* operational carriers and only one battleship, *Washington* (BB-56), but was too distant to engage. The carrier, *Enterprise* (CV-6), and battleship, *South Dakota* (BB-57), had taken bomb damage at the Battle of Santa Cruz and were repairing at Noumea.

The U.S. only had available : 2 heavy cruisers *San Francisco* (CA-38) and *Portland* (CA-33) ; 1 light cruiser *Helena* (CL-50) ; 2 antiaircraft light cruisers, *Atlanta* (CLaa-51) and *Juneau* (CLaa-52) ; and eight destroyers.

Both sides were in the process of sending full Army divisions to Guadalcanal. The U.S. arrived on Nov 12 and unloaded troops and most of their supplies before bombers damaged *San Francisco* and *Buchanan* (DD-484) was hit by friendly fire and the convoy withdrew. The Japanese sent a bombardment force of two battleships attended by a light cruiser and 8 destroyers, intending to take out Henderson Field to allow their troops to land on Nov 13 without being attacked by air. The U.S. cruisers and destroyers engaged the Japanese battleships that night.

The Japanese did not expect the U.S. force that night and the big guns were loaded with high explosives shells. The Americans maneuvered by radar to bring all guns to bear but were sighted by Japanese eyes and their guns were reloaded with armor-piercing shells. Suddenly a Japanese searchlight illuminated the lead cruiser. *Atlanta* fired her 16- 5" guns but received 14" shells and two torpedoes. *Cushing* and *Laffey* rushed to her aid and suffered a similar fate. *San Francisco* took 14" shells. *Portland* and *Juneau* each took a torpedo. A torpedo split *Barton* in two and *Monssen* took shellfire. After 24 minutes of confused firing, both sides retired. The escaping American ships were sighted by submarine *I-26* and the already damaged *Juneau* exploded and was gone when the smoke cleared taking the Five Sullivan Brothers with her. That night the U.S. lost six ships and 1,000 men including two admirals. The Japanese lost two destroyers, *Akatsuki* and *Yudachi*, but they had not hit the airfields.

Next morning, Marine fliers from the undamaged Henderson Field found a battleship slowed with engine damage and attacked with bombs and torpedoes. Still damaged *Enterprise* arrived to launch her aircraft, which landed on Guadalcanal when she retired. AAF B-17s participated. Her admiral ordered Japanese battleship ***Hiei*** scuttled.

The Japanese troop convoy of 11 ships and 12 destroyers was delayed for 24 hours to allow a second attempt to bombard Henderson Field, this time by heavy cruisers.

With the American fleet sunk or gone, two heavy cruisers were able to fire 1,000 rounds to disrupt the American airfield. Next morning, scout planes found the retiring Japanese. American planes attacked. ***Kinugasa*** was sunk and three more heavy cruisers were damaged. Meanwhile, the troop convoy of transports and destroyers was sighted. The planes shifted their attention. By nightfall **seven transports** had been sunk.

The Japanese rushed their remaining battleship and five cruisers escorted by nine destroyers to bombard the airfield to protect the landing.

The two American battleships with four destroyers arrived in time to meet the Japanese force. The two battle-wagons opened fire and the two sets of van destroyers engaged. Just repaired *South Dakota* (BB-57) had an electric fault that took her guns out of an early part of the battle and she took 42 heavy hits. *Washington's* (BB-56) 16" shells pelted the Japanese battleship. The destroyer fight saw **two Japanese destroyers** and **three U.S. destroyers** sunk. Her admiral ordered battleship ***Kirishima*** abandoned and sunk by torpedoes.

Meanwhile, the remaining **four Japanese transports** were run aground to most expeditiously unload 2,000 surviving troops, but their equipment was destroyed with the dawn. The Japanese now had 32,000 troops on Guadalcanal, but their Navy had to give up on support. The Tokyo Express of destroyers continued to run at night to bring in supplies, but only could bring in enough to sustain the troops; they could not bring any more reinforcements. The American airfields on Guadalcanal had survived and continued to grow. It took another two months for the Japanese Army to give up and withdraw. From this point onward, the Japanese could only retreat.

- At the Coral Sea, the Japanese advance was checked.
- At Midway, the odds were evened a bit.
- At Guadalcanal, the Japanese retreat began.

USS Washington (BB-56) 9- 16" guns

USS South Dakota (BB-57)

IJN Battleship Hiei 8- 14" guns

IJN Battleship Kirishima

Washington Naval Disarmament Treaty.

A Naval Disarmament Conference was held in Washington in 1921 with United States, Great Britain, Japan, France, Italy Belgium, Netherlands, and Portugal. Having fought the war to end all wars, countries were impoverished and needed to reduce ongoing government expense to dedicate funds to rebuild their cities. The general agreement was to reduce battleships to the level of the defeated counties. For the United States, the result was to decommission 39 battleships. The entire USN inventory of obsolete pre-dreadnought battleships (25) and those then building (7) were scrapped; two were converted to aircraft carriers, three converted to non-combat types, and two sold to Greece -- things the government wanted to do anyway. A treaty was signed in 1923.

A second conference was held in London in 1930, Italy did not participate, Japan soon felt a need for new construction without limitation and dropped out in 1936. So the followup conference on cruiser-sized ships was essentially self-imposition of limitations on the Allies. The US Navy scrapped 11 obsolete cruisers (coal burning and some with auxiliary sails); all 19 new cruisers built in the 1930s were limited in weight to 10,000 tons and arms to 6" guns for light cruisers and 8" guns for heavy. The USN built eight thin-skinned scout cruisers in this period and renamed as heavy cruisers because they carried 8" guns. In this same period, Japan built twelve heavies of 13,000 tons.

The U.S. did not commission any new battleships before the start of the next war and was unprepared with only 2/3 of authorized treaty strength. The limit on battleship size was 35,000 tons with 16" guns. Germany was not a naval power after WWI so was not invited and forbidden to have any battleships by the Treaty of Versailles. Hitler went ahead anyway and built the largest battleships the world had ever seen, *Bismarck,* 50,000 tons, 1940, and *Tirpitz,* 52,600 tons, Feb 1941. Meanwhile, Japan secretly had started construction of three battleships at 64,000 tons with 18" guns, *Yamato,* Dec 1941, and *Musashi,* 1942. The largest USN battleship was *Tennessee (BB-43),* 33,700 tons, 1920. Even the Iowa class battleship built during the War was only 45,000 tons, Feb 1943.

No U.S. capital ship authorized after Pearl Harbor was completed in time to participate in the war. By the time of the Battle of Leyte Gulf, Japan had fewer naval ships than the Allied forces had aircraft carriers, underscoring the disparity of industrial strength towards the end of the war.

Dud Torpedoes

The USN had a torpedo problem from the beginning of the war through mid-1943. For a year and a half, sailors put their lives on the line to make torpedo attacks with faulty equipment. Asiatic Fleet submarines made 136 attacks, firing 300 torpedoes in the first four months sinking only ten ships. If the duds had been able to effectively damage or destroy their targets, the Japanese expansion may have been more contained and the long path to ending the war may have been shorter and achieved earlier.

There were at least three problems with the Mark 14 submarine-launched torpedo.

1. (a.) **Running depth.** Warheads were **heavier** than test heads. The wartime torpedo ran with a head down trim. The shore establishment eventually acknowledged a four-foot error in running depths. Fleet tests in Australia found an 11-foot depth error. It was not until Aug'42 that a compromise of 10-feet was agreed and a trim repair kit was issued to the fleet later in that year.

(b.) The **depth sensor** was designed for a slower running torpedo. The pressure gradient over the torpedo surface at higher speeds gave the wrong feedback. The sensor was later relocated to a neutral position.

2. The **magnetic exploder** was designed in the northern latitudes and did not work as well at the equator. The British and Germans had already disabled their magnetic exploders before the USN ordered theirs disabled 24June43. ComSubSWPac had participated in the development of the magnetic exploder, knew the principle was sound, and resisted disablement until Dec'43.

3. The conventional **contact exploder** was designed for the earlier, slower, 33-knot, Mk 13 torpedo. The newer, faster, 46-knot, Mk 14 torpedo had higher inertial impacts that would cause the cross-wise mounted firing pin to miss the exploder cap.

How could these problems have gone unknown?

1. Depression era **economics** found a torpedo to be expensive to the fleet budget. Therefore torpedoes were not tested to destruction; they were fired against soft targets with floatable warheads so they could be salvaged for reuse.

2. The magnetic exploder was kept **secret** and had a complicated technique of attachment to the installed contact and was not issued to the fleet until needed for wartime patrol. With minimum testing and little fleet experience, the problems were unknown. The effects were failure to explode or to explode prematurely.

3. A **single source** of development, Naval Torpedo Station, Newport, RI, met reports of torpedo problems with disbelief that there could be anything wrong with what they considered were good designs when tested fine under controlled conditions. The problems first had to be proven by fleet units before the resources of the NTS were brought to bear.

4. Multiple problems take longer to find then does a single problem. Once a specific problem was uncovered, it could be quickly fixed. Only when the first problem was solved, could the nature of the succeeding problems be addressed, each in **sequence**.

    (1) Ref : Torpedo Scandal of 1942-1943 : from the October 1996 issue of *THE SUBMARINE REVIEW*.

Related Torpedoes.
Mark 10 Submarine Torpedo.

The older, slower, non-magnetic equipped torpedoes used in the older **S-boats** had one less of these problems. An error in its running depth was corrected by BuOrd just after Pearl Harbor ; it had run 4 feet too deep. It also had its share of premature detonation, erratic running, and other problems. Some subs and torpedoes were making hits, therefore problems were assumed to be human errors. Many good skippers were relieved for not sinking ships when it was the torpedo's fault.

Mark 13 Air Launched Torpedo.

This 2,200-pound weapon had been introduced in 1938, had not been adequately tested, but was the only one available. Its use required release at less than 60 feet above water and at slow airspeed. Over a year after the Battle of Midway and after some improvements, the Bureau of Ordinance ran a test with over 100 torpedoes and found only 31 percent gave a satisfactory run. Add to this -- exploder problems and erratic launches under fire against evading targets -- it is no wonder that dive bombing became the attack method of choice.

    (2) -- "Black Cat Raiders" by R. C. Knott
    The Mk 13 aerial torpedo design dated from the early 1930s. It was stubby.
    It's warhead originally carried 400 pounds of TNT, later 600 pounds of Torpex, that made it nose-heavy. Fleet squadrons were told to limit drops to 50 feet at no more than 110 knots from a range of approximately 800 yards. Performance was so bad (until improved in 1944) as to make American torpedo-plane attacks well nigh futile.

    (3) "Unknown Battle of Midway" by Alvin Kernan.

Air Attacks Made
- Lea   -- 13 launched, one hit, in harbor.
- Tulagi   -- 22 launched, one hit, in harbor.
- Shoho   -- 24 launched, 7 hits, already damaged by bombs.
- Shokaku -- 20 launched, no hits.
- Midway -- 51 torpedoes carried, no hits, 43 planes lost.
- Suzuya   -- 6 torpedoes, improved Mod 2, three reported explosions. Japanese reported no hits.

Tactical Problem
The torpedo was so slow and short range that a warship could outrun one of our torpedoes when not launched from ahead. The TBD Devastator torpedo bomber was too slow to get ahead of a turning ship. The torpedo made a clear wake allowing ships to steer away from its path. Some were observed in clear water on stationary targets to go under the enemy ship and detonate on the other side. Slow airspeed allowed fighters to easily catch the torpedo bombers and shoot them down and a required approach to 800 yards allowed hits from all calibers of ship-borne antiaircraft guns. The model 2 was introduced after Midway : many reported hits were actually premature explosions, according to Japanese records. Therefore many reported hits and sinkings were false.

Further improvements in 1943 allowed air drops at higher speeds and altitudes and resulted in 40% hit rates. The six engagements listed above with 136 torpedoes scored only nine hits.

Mark 15 Destroyer Launched Torpedo
This had all the same mechanisms as the Mk 14. It was just slightly larger (longer) version and should have had all of the same problems. The method of deployment and analysis of hits, misses, and duds between the submarine and surface fleet differed. A measured attack by a submarine would cause immediate note of a failure and trigger an analysis of the cause. A destroyer launched torpedo attack was in the face of greater excitement, often at night ; failures to hit while engaged in surface combat were not able to be immediately analyzed. After the battle, torpedo failures were simply attributed to misses by the crew. The poor performance of destroyer torpedo attacks can take on a new perspective, and need not be completely attributed to problems with tactics. Witness the failure to scuttle *Hornet* with nine destroyer torpedoes.

Shortage
The torpedo station at Cavite Naval Station (Manila) that was destroyed on 10 Dec included 2/3 of Asiatic Fleet torpedoes. This created a shortage such that many targets that would have been attacked with four torpedoes were targeted with only two. Even so, more torpedoes were fired in the first year than reached the fleet -- production was fouled up, too. Some subs were reduced to carrying half loads of torpedoes and had to take mines instead.

More.

Torpedo tubes were not reliable. Of the nine torpedoes launch attempts by subs at Midway, two failed to leave the tubes. Only one of the seven launched made a hit and it broke apart and was used as a life raft by Japanese survivors of the dive bombing attack. An unacceptably large number of torpedoes "prematured", that is, exploded on the way to the target. Because evading submarines could not see the results, any explosion heard near the target was assumed to be a hit. Later, when the submarine surfaced and found no ship, the assumption was that it had sunk, where it may have simply run away when warned by a premature. Credited sinkings were more than 50% greater than post-war confirmations and tonnage claimed was more than double that confirmed from Japanese records.

Some sub skippers, suspicious of the torpedoes, set the depth shallower and made hits which confused the results, others deactivated the magnetic exploder to further confuse the statistics. Orders were specifically reissued from headquarters to set torpedoes to run deep and to use the magnetic exploder. This was because the US torpedo was fairly light and not able to penetrate the side armor of a warship or sink a large merchantman, whereas a magnetically triggered explosion under the keel would break the back of a ship -- if it worked.

The direction control could lock "hard over" so that the torpedo circled, coming back at the sub that fired it. *Tullibee* (SS-284) and *Tang* (SS-306) were sunk by themselves in this fashion. Others were able to dive under the returning torpedo. *An extreme example of "friendly fire".*

Summary

In the first four months of the war, the 27 subs that comprised the Asiatic fleet sank only ten ships. Most skippers were replaced with more aggressive men. Yet problems continued. On 1 August 1942, BuOrd finally conceded the Mark 14 ran deep. On 9 April 1943 *Tunny* (SS-282) found herself in an ideal position to attack aircraft carriers *Hiyo, Junyo,* and *Taiyo.* From only 880 yards (perfect, close range), she fired all ten tubes, hearing all four stern shots and three of the bow's six explode. No enemy carrier was seen to diminish its speed. Later, intelligence reported each of the seven explosions had been premature ; the torpedoes had run true but the magnetic feature had fired them too early. Finally, in July 1943, Admiral Lockwood ordered his boats to deactivate the magnetic influence exploder." "Duds" -- torpedoes heard to hit but not explode -- were addressed in September 1943 when the first torpedoes with new contact pistols were finally sent to war. For fully half of the war, submariners, airmen, and destroyermen had risked their lives with faulty equipment.

Judgment

"It is sadly true that each modern torpedo type sent to war by the United States Navy was defective. ... The failure to test this crucial weapon prior to hostilities created the greatest technological failure in the history of American military." -- *"Fire in the Sky"* by Eric Bergerud.

Submarine :

	US Mark 10	US Mark XIV	Japanese Type 95
Diameter :	21 inch	21 inch	21 inch
Range :	1-3/4 miles	2-1/4 miles	5 miles
Speed :	36 knots	46 knots	50 knots
Explosive :	500 pounds	643 pounds	893 pounds

Surface :

	US Mark 15	Type 93 Long Lance
Diameter :	21 in	24 inch
Range :	7-1/2 miles	25 miles
Speed :	45 knots	54 knots
Explosive :	825 pounds	1,058 pounds

Japanese Torpedos

The Japanese submarine torpedo, Type 95, a down-sized Long Lance, was the same 21" diameter of the U.S. types. The Long Lance, or type 93 torpedo, was 24" dia. used with telling effect by cruisers and destroyers. Top speed was 54 knots (60 mph), maximum range 50,000 yards (25 miles), with a 1,058-pound (half ton) warhead. Pure oxygen fired the fuel giving high performance and little surface wake. In the Battle in the Java Sea, Feb 1942, the ABDA warships that closed to gun range thought they were out of torpedo range, were surprised and assumed they had hit mines or that submarines were in the area.

Japanese Kaiten Torpedo. Late in the war, the man-guided Kaiten or "Heaven Shaker" suicide torpedo was developed from an extended Type 93 of 8 tons with 3,000 pounds of warhead with a range of 43 miles. Beginning in November 1944, five submarines set out into the Pacific on Operation Kongo each with 4 Kaitens strapped on deck. They sank two ships, and damaged several more, causing concern among the U.S. forces, but the operation was generally unsuccessful costing 187 American lives vs. 948 Japanese sailors, mostly from mother subs that were sunk.

Battle of Tassafaronga -- 30 Nov 42
After the Big Battle.
U.S. Cruiser Force Destroyed

The Japanese tried to replenish their troops on Guadalcanal by fast destroyers on moonless nights. The destroyers hid in Bougainville during the day, made a run to deliver their supplies in the dark and returned to hiding by daylight. The technique was to tie barrels together and drop them overboard where they were retrieved by swimmers and hidden before the morning when an attack by Marine planes was assured.

The USN developed a plan to ambush such convoys. When a convoy was reported, a task force of cruisers and destroyers would wait in their path, detect the destroyer convoy by radar, launch a silent torpedo attack, and, after the torpedoes had done their work, attack the surviving ships with overwhelming gunfire.

A convoy was reported heading for Guadalcanal for the night of 30Nov42 and TF 67 sailed with four heavy cruisers, one light cruiser, and six destroyers to intercept in Savo Sound off Lunga Roads. The Japanese convoy of eight DD sailed in line astern; six were encumbered by drums of supplies, two were scouts and escorts.

Radar aboard the flagship detected the convoy at eleven miles and the U.S. task force turned to parallel the enemy. The Japanese sailed close to the coast where they could not be tracked clearly and were out of effective torpedo range. While commanders discussed the situation in which land-based planes failed to appear to drop flares, the convoy passed its closest point of approach and twenty torpedoes were finally fired as the distance lengthened. Star shells were fired before the torpedoes completed their runs. The Japanese replied with a barrage of their excellent, long-range and accurate torpedoes at the American gun flashes. The newly commissioned, Japanese scout destroyer, ***Takanami,*** reversed course firing incessantly, drawing the concentrated fire of the American ships while the convoy released their supplies, reversed course and launched a second salvo of torpedoes. Four U.S. cruisers took six torpedo hits and the remaining seven Japanese destroyers made good their escape. Heavy cruiser ***Northampton*** sank while *Minneapolis, New Orleans,* and *Pensacola* were badly damaged. Only the light cruiser, *Honolulu,* and the six destroyers escaped damage. However, this battle helped end the attempt to resupply by the "Tokyo Express."

Japanese Submarines

The Japanese also used submarines to ferry supplies to their troops on Guadalcanal. The U.S. had air superiority and underwater supply was deemed necessary. As many as 38 subs were stripped of arms and used as transports at the cost of twenty lost. Consider that the few remaining submarines that were engaged in an attack role sank *Yorktown* (CV-2), *Wasp* (CV-7) and damaged *Saratoga* (CV-3), *Hornet* (CV-8) and *North Carolina* (BB-55) plus other warships and merchantmen.

RO-33 – Coastal Submarine

I-52 – Long Range Fleet Submarine

I-25 had a range of over 14,000 miles, carried a cruiser sized, 5.5" deck gun that fired on Vancouver Lighthouse and Ft. Stevens, OR. Seventeen 21" torpedoes that sank SS *Camden* and *SS Larry Doheny* off California. And an E14Y Glen seaplane that twice bombed Oregon with two thermite bombs each time, trying to set forest fires.

Eleven I-class submarines carried reconnaissance seaplanes in 1941-42. An equal number could carry mini-subs. Others could lay mines. All carried cruiser sized deck guns.

Midget Submarine recovered from Pearl Harbor attack.

Mini-subs were carried to their target area by fleet subs and used to penetrate harbors.

Where are we and How did we get here?

Japan was the most advanced nation in Asia. They thought they should rule and set about to do it. They were in the process of conquering China, but, because Japan is a resource-poor island, they needed oil and mineral imports from the touchy Americans and from the Dutch East Indies. (The Southern Resource Area was their name for it.) The Dutch and the British were occupied with some survival issues in Europe. Japan's leaders thought there was no need for difficult negotiations, just take what was needed for national success and settle up later. The Axis were succeeding in Europe, so Japan joined the Tripartite Pact. This strained relations to near breaking point with the USA, a resource provider to Japan of gasoline and scrap iron for the military. By joining the Axis Japan became a belligerent of our Allies. Go back several years, the U.S. and most of the world looked favorably upon the advanced Japan being a guide to backward China, -- until the Nanking Massacre of up to 300,000 people. World opinion towards Japan fell. The Pact with the Axis was too much.

In parallel, the U.S. sided with England against Nazi, The USN was in combat in the Atlantic as "all steps short of war." One-third of the U.S. Pacific fleet had been sent to the Atlantic where they fought the Germans by replacing the Royal Navy in the western half of the Atlantic. We escorted convoys, attacked U-boats, radioed for English ships to come take German merchant ships we found. An American pilot flying an American plane given to England spotted the *Bismarck*. U.S. battleships, an aircraft carrier, and merchant ships made up a convoy transporting 20,000 British troops, from Halifax to the Far East in November ; the USN escort turned back Dec 7. The British troops landed at Singapore while Japanese shelled the U.S. troop transports. The ships escaped, the troops marched into prison camps. For the U.S. to have continued sending resources to Japan after they joined the Axis Pact would have been to directly aid our enemy's war effort. The shut-off of American oil and iron emphasized the importance of the Dutch East Indies to the Japanese.

One can, with a few moments thought, easily list the steps required by Japan. England held the fortress at Singapore, The U.S. had the Philippines. These had to be neutralized before taking over Java, Borneo, etc. The Army would take British Malaya (launched from IndoChina, provided by Vichy) by the back door -- Singapore was defended from the seaward side, the land approach was open. The rest of SE Asia was not militarized and they could walk through one country to the next without opposition. The Australian's were busy in North Africa. The British were engaged with their tussle in Europe,

Till they met Fletcher

so their only ally, the USA, had to be dealt with. The U.S. presence in the Territory of the Philippine Islands had to be addressed with force, but they were weak and unprepared. The real concern was the obvious expectation of the USN to sail across the Pacific to defend the Philippine Islands.

Well, what to do? In the Russo-Japanese war of forty years earlier, the Japanese had launched a surprise attack. Let's try it again ; this time to take out the U.S. Pacific Fleet. Other points of interest to Japan : within a month of the fall of France in the prior year, FDR authorized a Two Ocean Navy of 210 new ships including 11 aircraft carriers and 7 battleships. The first of these would commission in early 1942. MacArthur was preparing for war in April 1942. The Japanese Navy was stronger in the Pacific in 1941 than the combination of all other nations. The Japanese Army and Air Force had four years of combat experience and weapons development from warfare in China. The Japanese Navy had ten battleships and ten aircraft carriers. The U.S. had in the Pacific eight battleships and three aircraft carriers, you get the idea. After Pearl Harbor, we had zero battleships in service in the Pacific. The goal of Pearl Harbor was not to war with the US, but to discourage the U.S. from going to war with Japan.

The Japanese thought they could become the only Asian power and with the whole Pacific and Indian Oceans as buffers from retaliation. Their thinking was that after they got the Southern Resource Area and a ring of defensive positions, they would favorably consider a request by the world for peace. They would agree, and get on with their real goal -- the conquest of China.

Subsequent events proved they were almost right. They defeated the U.S. Battle Fleet, the U.S. Asiatic fleet, the British Far East Fleets (two of them), and the Royal Dutch Navy ; they attacked from Hawaii to India, from Alaska to Australia, without losing a ship larger than a submarine or auxiliary destroyer. But, politically, they expected the world, (e.g., isolationist USA) to be reasonable and accept the new occupations by Japan and negotiate a peace because this was the only logical thing to do.

But Fletcher slowed them down and sank six of their carriers. Battleships turned out to be of little value in the air age, The Axis built ten aircraft carriers during the war ; the U.S. alone built 119. Once mobilized, the USA was overwhelming. Contrast this with the call in Congress of a few years earlier, "Schools, not ships."

But this was in the future. As the middle of 1942 arrived. the Japanese had accomplished their goal of defeating all navies and established a defensive ring. That had been so easy that they expanded their goals until they bumped into Fletcher. They had overextended themselves and had not the industrial base to recover from the loses that Fletcher had inflicted upon them The future could hold only retreat.

Andres del Castillo, author of *The Days of Fletcher*, says Fletcher won the Atlantic war, too, by allowing FDR to continue with the "Germany First" policy. If Fletcher had failed, then the Lend Lease aid to England and Russia and the invasion of North Africa and the underbelly of Europe would have been delayed and made more expensive in men and material. What makes Admiral Fletcher's success most significant, as the strong arm of Nimitz, is that they did it with the survivors of an unappreciated, peacetime Navy against a larger, better armed, and battle trained enemy.

Fleet Admiral Chester W. Nimitz signs for the United States.

Part 3 – Reference

Grand Strategy

Grand Strategy is determined by the heads of state representing the recommendations of their military and political staffs. All agreements were compromises. The principle concern of such international efforts, or Grand Strategy, was Germany. Japan was left as the concern of the United States and the Grand Strategy for the Pacific War was concerned primarily with providing only the minimum of U.S. resources that could be allocated to the Pacific as a holding action without taking away from the (real) war against Germany.

1939.
Sep 1 . Germany attacks Poland ; war declared 48 hours later.
Sep 4 . U.S. extends Atlantic defense zone to 300 miles.
Oct 28. Italy attacks Greece.
Nov 4. U.S. repeals arms embargo, allows Cash and Carry of military supplies in foreign ships. U.S. arms industry brings nation out of depression.
Nov 30. Russia invades Finland. Finns win Winter War, lost in the Spring.

1940.
January. Authorized sale of late-model U.S. military aircraft.
April 9. Germany seizes Norway and Denmark.
May 7 . Churchill forms a government.
May 7 . Pacific fleet to remain at Pearl Harbor as a sign to deter Japan.
May 10. Germany invades Holland and Belgium.
June 03. British evacuates from continent at Dunkirk. France falls on 17th.
July 10. Aerial Battle of Britain starts.
July 19. U.S. authorize 2-Ocean Navy, to build 7BB, 11CV, 33CL, 115DD, 42SS.
July 25. Embargo of metal and oil to Japan.
July 26. MacArthur appointed commander of all U.S. forces in Philippines.
Sep 16. Selective Service Act, "the draft".
Sep 17. German invasion of England canceled
Sep 27. Japan, German and Italy in Tripartite Pact forms Axis.
Oct 12. Japanese decision to go to war to obtain resources.

1941.
Jan 6 . Lend-Lease introduced to Congress.
American-British Conference Jan 29-March 27, 1941 . ABC-1 staff discussions in Washington, D.C.
    Europe first.
    Save Australia, New Zealand, and Singapore.
    Churchill from May 15, 1940, offered/wanted USN in Singapore.
    U.S. not approve Singapore. Separately to built up Philippines.
    U.S. commanders to observe in UK : MGen Chaney, RAdm Ghormley.
    British Joint Staff Mission to Washington, D.C.
Feb 17. Africa Corp lands in Libya to replace defeated Italians.

March. U.S. ships to report sightings of U-boats in Western Atlantic.
Mar 11. Lend-Lease approved by Congress.
Mar 29. **ABC-2.** Air recommendations issued.
   US aircraft production till Nov to RAF. Equal shares when US enters war.
Spring. Rainbow was a set of war plans developed 1939-41, each based on various possibilities. Rainbow 5 met conditions agreed on in ABC-1. Approved by Army-Navy Board May 14, Secretaries Army and Navy on June 2nd. U.S. to be ready to commit to plan Sept 1 or later. Specifies sequence, locations and numbers of U.S. troops and planes for garrisoning overseas bases for the first six months after mobilization for war, M-Day.
Apr 7 . Quarter of Pacific Fleet ordered to Atlantic : 3BB,1CV,4CL,18DD,3AO.
April . U.S. Army position : Able to occupy Iceland, Azores, Cape Verde Island, Caribbean or NE Brazil subject to availability of shipping. Not ready for full war. But, better to enter war before fully ready to save British Isles physically or politically.
   Lend-Lease extended to China.
May 27. "Sink the *Bismarck*". Unlimited national emergency declared by U.S.
June 10. Orders to survey locations for U.S. bases in England.
June 22. German invasion of USSR. End of Aerial Battle of Britain.
June 24. Japan pressures Vichy for Indochina, start build-up of invasion troops.
July 01. President orders Marines to Iceland. Arrive July 7.
   Plenty of drafted troops, but untrained and unequipped.
July 29. Japan occupies (Vichy) French Indochina.
Argentia Conference Aug 11-14, Newfoundland talks FDR and WSC.
   British plan : U.S. to North Africa and Mideast;
   Armored attack with partisan support on Europe.
   Joint British, Dutch, U.S. warning to Japan.
   U.S. listened, made no commitments and responded later.
Aug 14. **Atlantic Charter** -- political statement on conclusion of talks.
Aug 17. U.S. note to Japan : continued military action will necessitate actions in defense of U.S. national security. *A threat?*
Aug . U.S. Lend-Lease Mission to China.
Sept . U.S. Army internal response. U.S. keeps neutral, supplies Britain, trains and equips U.S. troops.
   Victory Program -- plan for war material needs.
   Strategic Concept plans for 8.8 million men to attack Germany by July 1, 1943.
Sept 5. U.S. Army sails to Iceland -- disrupts training programs to get qualified troops.

Oct 1 . US-USSR survey of material needs.

Oct 3. *Japanese Pearl Harbor Plan.*
Oct 5. Naval Conference between U.S. and British commanders at Singapore.
Oct . Lend-Lease Mission to Middle East.
Oct 16. *IJN Pearl Harbor Force training in Kuriles*
Nov 18. Five mother subs, each with midget sub, depart Kure for Pearl Harbor.
Nov 25. *Sortie of a Japanese invasion fleet towards Malaya.*
Nov 27. *Sortie Mobile Fleet of 6 carriers into North Pacific.*
Nov 30. Lend-Lease extended to USSR.
Dec 7 . Invasion of Malaya. Attack on **Pearl Harbor**. Invasion of Guam. Raids on Wake, Midway, Philippines.
Dec 8-11. Declarations of war all round.

Arcadia Conference. Washington, D.C. Dec 17 - Jan 14.
American-British Conversations - Dec 22. Churchill to D.C. for full staff conferences.
British plan to land in Europe in summer of 1943.
    Minimum safeguards in other theaters.
    Shipping limits U.S. troop deployment to only one of the three areas.
    Pacific Tasks 1942 :
        1. Maintain Australia, New Zealand, India and supply China.
        2. Hold Hawaii and Alaska ; Singapore, East Indies Barrier and Philippines ; Rangoon and route to China.
British proposal to invade French North Africa.
    Seeks U.S. to land in Morocco.
    U.S. says there is a crisis in Pacific, postpone N. Africa until shipping becomes available.
Dec 25. Hong Kong falls.

> Definitions: Combined = International, Allied
>   Joint = Inter-service, American

1942.
Jan 1 . Declaration of the United Nations -- 26 signers.
Jan 2 . Manila falls. Japanese landings in Borneo and Celebes.
Jan 15. ABDA Command established of American, British, Dutch, and Australian military forces under General Sir Archibald Wavell
    First U.S. troops sail for North Ireland.
Feb 15. Allies defeated in Battle of Java Sea; Singapore falls.
    Allied Naval and Air forces are out of Western Pacific.
Mar 1 . ABDA command disbanded.
Apr 1 . U.S. is planning for invasion of Europe in Spring '43.
May 8. Corregidor falls. Allied armies are out of Western Pacific.
May-June. Fletcher stops Japanese advance in South and Central Pacific.
June . Brits convince U.S. to proceed with N. Africa in 1942; postpone Europe.
Aug . Iran is established as an entry point of Lend-Lease to Russia.
Aug 7 . Marines land on Guadalcanal.
Nov 7 . Allies invade North Africa : 140,000 British ; 150,000 U.S.
Nov 11. Germany occupies the rest of France.
Nov 15. Naval Battle of Guadalcanal sends both navies back to rebuild.
Nov 20. Soviets encircle Germans at Stalingrad.
    U.S. debates Central Pacific or South West Pacific strategy.

Reference

1943.
Casablanca Conference. Jan 14
Proposals : British wanted to continue taking all of N.Africa and open the Mediterranean to shipping.
Attack the soft underbelly of Europe.
U.S. wanted to build up forces in England for cross-channel attack on northern France in 1943.

Decided :
Europe : Weaken Germany ; Knock Italy out of war ; Take pressure off Russia : Bring Turkey into War against Germany.
England : Build up forces in England ; Delay cross-channel till Spring'44.
Air operations in Europe by priority : submarine construction, aircraft industry, transportation, oil plants, other war targets.
Mediterranean : Immediate plans to invade Sicily.
CBI : Postpone Burma till late '43.
Pacific : Increase allocation from 15% to 30% of war material.

Jan 24. End conference with "unconditional surrender" statement.

Trident Conference. D.C. May 12.
Proposals :
U.S. wants earliest possible cross-channel invasion.
British considered "possible" only if Germany weakened significantly before invasion ; wanted to continue attacks throughout Med, specifically to knock Italy out of war by capture of Rome.
U.S. wanted to stop with boot and heel of Italy for air bases and no Mediterranean operations east of Italy.
To get USSR into Far East War after surrender of Germany.
U.S. wanted Brits and China to open supply line to China.
British saw China of little value in war and refused a land war in Burma. China would not move unless RN attacked Rangoon.
China wanted increased air attacks on Japan from China bases.
U.S. saw this as backwards, need supply lines first.

Compromises :
Europe : Fix date for cross-channel attack for May 1, 1944.
Mediterranean : Limit forces in Med to 4 U.S. and 3 Brit divisions.
Any surplus resources, such as by delay in cross-channel, to go to Pacific, not Mediterranean.
China : U.S. to double, then, triple cargo by air over "hump" giving priority to air war.

Pacific : All agreed with U.S. plans in the following steps:
1 . British and China open lines thru northern Burma. U.S into the Celebes Sea between New Guinea and Borneo.
2 . U.S. to retake Philippines ; British to raid Straights of Malacca.
3 . China to attack Hong Kong with U.S. covering that part of South China Sea from Philippines.
4 . U.S. to capture bases for airfields within range of Japan.
5 . Air offensive against Japan proper.
6 . Invasion of Japan.

Algiers Conference May 29-June 3.
Churchill tries to convince Marshall and Eisenhower to take Rome in 1943. Ike agreed that if Sicily was easy, he would maintain momentum by taking a bridgehead on Italy, so long as it did not take from the buildup for cross-channel.

June 11. After two weeks of heavy air attack, Pantelleria surrenders.
July 10 - Aug 17. Invasion of Sicily.

Quadrant Conference Aug 17-24 in Quebec.
Proposals : British had a list of requirements before going ahead with "Overlord", the cross-channel invasion, and did not want to take troops from Mediterranean. U.S. said either cross-channel as planned or U.S. efforts will go to the Pacific.

British wanted Italy well north of Rome ; for French troops to land in S. France ; and half again as large D-Day landing force. The U.S. only wanted the airfields in Southern Italy and did not have enough landing craft for both Southern France and D-Day cross-channel.

U.S. wanted to divert Med shipping to Pacific to attack Rabaul and a plan to redirect resources from Europe to Pacific on defeat of Germany, expected in 1944. Brits saw no need for continued advance north from New Guinea.

U.S. wanted British to take northern Burma by Nov'44 giving a supply line from India to China to support B-29 bases in China to attack Japan. Then to begin on southern Burma and Singapore. British wanted to postpone Singapore till March'45 and southern Burma till Nov'46.

Decisions :
1. Mediterranean : constant pressure on Italy.
2 . Cross-Channel : increase landing force, keep the date.
3 . Pacific : U.S. thrusts thru Central and from S.W. Pacific.
4 . Burma : Burma road thru north to begin Feb'44.
Southeast Asia Command established.

Reference

Moscow Conference. October 19-30 of Foreign Ministers
   Discuss UN, Italy, Austria, atrocities
Cairo Conference. Nov 23-26 – British and U.S. meet with China on way to Tehran to meet with Russians.
Tehran Conference. Nov 28-Dec 1.
   To transfer troops in Italy to invade south of France.
   Simultaneous offensive by Russians
   Soviet entry in war on Japan after defeat of Germany.
Cairo, again. Dec 4-6, 1943
   Chiang demands $1 billion in gold to stay in war (his concern was communists).
   Not use Chinese troops in Burma.
   Postpone Burma to allow landing craft for cross-channel.

Year-end U.S. military troop strength : In Britain 800,000 men (40% air); Mediterranean 600,000 ; Pacific 700,000 (25% air), 800,000 Navy.
   Delay and uncertainty in cross-channel had allowed buildup in Pacific faster than original plan. The Navy had a clearer picture of the Pacific needs.
   Shipping is about equally split between Atlantic and Pacific ; distances were greater in the Pacific, so fewer round trips per ship.
   The shortage of LSTs that hampered European Planning was caused by concentration on building of destroyer escorts to combat U-boats. The DEs started arriving after the peak in U-boat activity (reduction caused by B-24s, ULTRA, CVEs) just at the time that the LSTs were needed in all three theaters.

1944.

Jan 22- May 23. Anzio, Italy. Establish a beachhead, but unable to breakout and join with frontline coming from the south.
   Debate over Burma Road, which requires Chinese army, vs. taking a port on the China coast which would be a major effort by Allies taking away from other areas.
Mar 11. Japanese incursion into India ends debate. Thrown back by July.
   C-47s taken from Overlord to airdrop munitions.
Apr 18. Anzio causes delay in invasion of southern France till after D-Day.
May . Japanese offensive into Southern China to eliminate U.S. airbases.
   U.S. debates next Pacific move between Formosa or Okinawa.
June 6. Cross-channel invasion, D-Day.
Jun 15. B-29s bomb Japan from China till Nov ; every drop of gas and bullet is delivered by air over the "hump".

Bretton Woods. New Hampshire. July. 1-22. 44-nations create the international monetary system

Dumbarton Oaks. D. C. Aug-Oct. Four-powers discussed United Nations, Security Council, and the veto.

Octagon Conference. second Quebec. Sept 13-16. Start on post-war issues.
International Monetary Fund created to stabilize currency and trade.
International Bank for Reconstruction which becomes World Bank.

Nov 24. Start of B-29 raids on Japan from Tinian, Marianas.
Japanese overrun U.S. air bases in China and Chinese army collapse.
U.S. downgrades future China efforts as a drain with no prospects.

1945.

Yalta Conference. Crimea. Feb 1-11. Post-war issues.
    1 . Set occupation zones in Europe.
    2 . USSR to enter Pacific War 90 days after defeat of Germany.
    3 . Soviets to get parts of Poland and Japan.
    4 . Free elections in Europe.

Apr 12. FDR dies ; Truman is President.

May 7 . **Germany surrenders.**

Potsdam Conference. outside Berlin. July 16 - Aug 2.
    1 . Allied control of defeated Germany.
    2 . Reparations.
    3 . Oder/Neisse Line giving part of Poland to Russia and part of Germany to Poland.
    4 . Russia joining the war in the Far East is confirmed.

Aug 6 . Hiroshima.

Aug 8 . USSR invades Manchuria per Potsdam agreement. Fight and occupy land area till Sept 2.

Aug 14. **Japan surrenders.**
U.S. stops fighting, aircraft jettison bombs and return.
Soviets continue to seize land till signing

Sept 2 . Signing of surrender in Tokyo Bay.

Why Concentrate on the Early Years ?

Everybody, except some Japanese students, knows of the Japanese expansionist attacks on Pearl Harbor, Thailand, Malaya, Philippines, Dutch East Indies, Ceylon, Australia and the United States. And know that the United States retaliated with overwhelming force to conclude the Pacific War that Japan had initiated to support her attempted conquest of mainland China that had begun over four years before. However, almost two years were required to mobilize American industry to provide the ships, planes, and trained men to achieve this victory. The great battles late in the war -- Mariana's Turkey Shoot, Leyte Gulf, Saipan, Iwo Jima, and Okinawa, and the bombings that finally terminated it all are well known, as is the expected end of the war by the invasion of the mainland with the mighty American Pacific forces augmented by air forces released from the European Theater.

What is little known is the early war, when a peace-time military, under-funded during the Great Depression, had to try to stop the advance of a trained, equipped and experienced Imperial Japanese Navy and Army. The Japanese swept all before them. The IJN traveled from Pearl Harbor to India, from Australia to Alaska, sweeping away the American battle fleet, the U.S. Asiatic Fleet, the British Far East Fleet, the Dutch fleet, with no loss to themselves. The professional core of the American military threw themselves into a failing defense without sufficient weapons in number or modernness.

The leadership in Washington cried out for instant victory from the very armed services they had only months before begrudged funding. The Japanese were superior in aircraft, in torpedoes, in aircraft carriers, in night fighting, and had suicidally great morale. The United States had only potential, if the war could wait until it was mustered, and some exceptional bravery. "They were expendable" in the early years. In 1942, the issue was in doubt. By the end of 1943 there was no doubt, the enemy just had to be convinced they had lost. Most of the battle deaths of the war occurred in proving this point. The period of Victory is well reported. The time of waiting, fighting as the underdog is crucial and not so well recorded. Hence this book.

Winning the Pacific War started at the Coral Sea where the Japanese were moving towards Australia (the Australian Army was in North Africa). An invasion force was moving to invest the coast of New Guinea from which they could suppress Australian participation in the war. They were met with a U. S. task force of two aircraft carriers, the Japanese had three. The Japanese troop transports were covered with a carrier and seven cruisers; they were met by an Australian force of three cruisers. For the first time the Japanese lost a capital ship and they retreated. The cost was three U.S. ships, but for the first time the Japanese had been thwarted in their advance.

This was followed a month later when the entire Japanese fleet (10 carriers, 11 battleships, total 116 surface ships the size of destroyers or larger plus 16 submarines) moving in on Midway was ambushed by the entire U.S. Pacific fleet (3 carriers, 0 battleships; total 44 surface warships destroyer or larger and 25 subs). The Japanese lost four big carriers, the U.S. lost one. The overwhelming Japanese capacity was reduced to simply being larger than the American numbers. Soon after, the U.S. met the Japanese for a six-month sea, air, and land battle at Guadalcanal that left both navies exhausted. Forty-eight warships were sunk there, an equal number on each side. The Japanese started the withdrawal that would continue step by step for two and a half more years. Every step was hard fought on both sides, but while Japan built ten new aircraft carriers, the U.S. built over one hundred. The issue was no longer in doubt.

There are four periods of the war from an American perspective.

>   The period of defeat --   Dec 7, 1941 to May 1942
>   The period of equalization -- Jun'42 to Dec 1942
>   The period of parity --   Jan'43 to Oct 1943
>   The period of winning --   Nov'43 to Aug 1945

We like to concentrate on the crucial period when we were able to stop, if not yet defeat the enemy. This is beginning to be called the "Days of Fletcher," because the first three battles between aircraft carriers were commanded by Frank Jack Fletcher in which he sank six enemy carriers with the loss of only two. The Japanese Navy was unstoppable until they met Fletcher. The marvel of his achievement is simply noted in that when he was replaced, his successors also lost two carriers, but in only six weeks and without sinking any enemy carriers. Fletcher had equalized the war, but it was not yet won.

This early period is almost unknown; Fletcher himself is almost unknown. The victories in grand battles later in the war are publicized because the numbers of people, ships and planes were great and the events were clear victories. There were more people to write about these actions. But consider that at the beginning, there were both fewer numbers and that many of the small number of heroes from early in the war did not survive to tell their tales. The admirals you have heard about -- Halsey, Spruance, Mitscher, Kinkaid -- led overwhelming forces against an enemy weakened by those early men.

Reference

To emphasize the point about later victories and the famous admirals -- the march across the Central Pacific began after a lull of almost a year after Guadalcanal and Fletcher had left the scene. The Gilberts campaign began in mid-November 1943 with invasions of Tarawa and Makin.

> "18-26Nov43. Occupation of the Gilbert Islands. Six heavy and five light carriers opened the campaign to capture the Gilberts with a 2-day air attack on airfields and defensive installations (18-19 Nov), covered the landings of Marines and Army troops on Tarawa and Makin Atolls (20 Nov) and on Abemama (21 Nov), and supported operations ashore (21-24 Nov). Eight escort carriers, operating with the Attack Forces, covered the approach of assault shipping (10-18 Nov), flew anti-submarine and combat air patrols, and close support missions on call (19-24 Nov)." -- *Naval Aviation Chronology in World War II*, Naval Historical Center.

Count them : 6 CV , 5 CVL , 8 CVE = 19 carriers ; plus 12 battleships.

By contrast, the Japanese started the war with ten carriers of all types and built or converted 14 more during the war (plus two BBV, with five carriers building at war-end) and Fletcher sank the four biggest in his bag of six while never having more than three carriers. When Spruance attacked the Gilberts there were only ten Japanese carriers in the whole Pacific, plus a training carrier. Japan had none near the Gilberts.

Spruance had twice as many carriers in his first operation than the enemy had available in *total* at home and deployed. Six times more than Fletcher ever had. And the U.S. had to confront none. The numbers did nothing but improve. By the end of the war, the U.S. had over 88 carriers available (96 commissioned during the war, 27 building) plus the British carriers (10 plus 33 lend lease). You have heard the names of those who commanded these large numbers – Halsey, Spruance, Mitscher, Kinkaid, . . . but not Fletcher.

Concerning Battleships -- Japan had twelve sometime in the whole war : two were sunk at Guadalcanal, two were converted to hybrid aircraft carriers that saw no service, one self-exploded in Hiroshima harbor, so that Japan had seven battleships available and five of these were older than anything the U.S. had in the Pacific. But two were new, super battleships, the largest in the world. The point is that Spruance's first operation in the Gilberts was supported with 12 U.S. battleships that outnumbered the entire Japanese battle fleet of seven. Japan had none near the Gilberts. The U.S. had one other battleship in the Pacific and three in the Atlantic. Fletcher never had more than one and that at only one of his battles.

War - The First Days

Dec 8, 1941, Monday.
* Japanese Imperial Rescript declaring a state of war between the Japanese Empire and the United States is issued in Tokyo. [Note : the "Fourteen Point message" delivered while the attack on Pearl Harbor is in progress merely declares an impasse in the ongoing diplomatic negotiations.]
* U.S. declares war on Japan. In his address to the nation, President Roosevelt describes December 7th, 1941, as "a date which will live in infamy."
* Britain, Canada, Costa Rica, Dominica, Haiti, Honduras, Nicaragua, Free France and Netherlands governments in exile declare war on Japan. China declares war on German and Italy.
* Thailand capitulates within a few hours of Japanese invasion.
* Japanese submarine *I-123* mines Balabac Strait, Philippine Islands, and *I-124* the entrance to Manila Bay.
* Striking Force, U.S. Asiatic Fleet departs Iloilo, Philippine.Islands. for Makassar Strait, Netherlands East Indies.
* Seaplane tender (destroyer) *William B. Preston* (AVD-7) is attacked by fighters and attack planes from Japanese carrier *Ryujo* in Davao Gulf, P.I.; *William B. Preston* escapes, but two PBYs she is tending are strafed and destroyed on the water.
* Japan interns U.S. Marines and nationals at Shanghai, Tientsin, and Chinwangtao, China. River gunboat *Wake* (PR-3) maintained at Shanghai as station ship and manned by a skeleton crew is seized by a Japanese Naval Landing Force boarding party after an attempt to scuttle fails.
* U.S. passenger liner **President Harrison**, en route to evacuate Marines from North China, is intentionally run aground at Sha Wai Shan, China, and is captured by the Japanese, refloated, repaired, and renamed *Kakko Maru* and later, *Kachidoki Maru*. Among the baggage awaiting shipment out of occupied China along with the North China Marines are the bones of Peking Man, which are never seen again. Their fate remains a mystery to this day.
* *Saratoga* (CV-3) departs San Diego for Pearl Harbor.
* *Wake*, the only U.S. Navy ship to surrender during World War II, is renamed *Tatara* and serves under the Rising Sun for the rest of the war. British river gunboat HMS **Peterel**, however, moored nearby in the Whangpoo River refuses demand to surrender and is sunk by gunfire from Japanese coast defense ship *Idzumo*. Five American-flag tugboats are seized by the Japanese at Shanghai.
* Japanese forces land on Batan Island, north of Luzon, P.I.
* Japanese forces move rapidly down the east coast of Malay Peninsula.
* RAF Hudsons bomb invasion shipping off Kota Bharu, Malaya, setting Army cargo ship *Awajisan Maru* afire.
* Japanese planes bomb Hong Kong, Singapore, and the Philippine Islands. Extensive damage is inflicted on USAAF aircraft at Clark Field, Luzon, P.I.

Reference 189

During Japanese bombing of shipping in Manila Bay, U.S. freighter ***Capillo*** is set afire and abandoned.
* Japanese naval land attack planes bomb Wake Island, inflicting heavy damage on airfield installations and F4F Wildcats. A U.S. four-plane patrol is out of position to deal with the incoming raid (there is no radar on Wake). Pan American Airways Martin 130 Philippine Clipper (being prepared for a scouting flight with an escort of two F4Fs when the attack comes), in the aftermath of the disaster precipitately evacuates Caucasian airline staff and passengers. Pan American's Chamorro employees and a government accountant are left behind.
* Japanese force sails from Kwajalein, in the Marshall Islands, to assault Wake Island.
* Japanese floatplanes bomb Guam, M.I., damaging minesweeper *Penguin* (AM-33) and auxiliary oiler *Robert L. Barnes* (AG-27). ***Penguin*** is abandoned and scuttled in deep water by her crew.

Dec 9, Tuesday.
* Japanese seize Tarawa and Makin, Gilbert Islands.
* Japanese submarines *RO-63, RO-64,* and *RO-68* bombard Howland and Baker Islands in the mistaken belief that American seaplane bases exist there.
* Transport *William Ward Burrows* (AP-6), en route to Wake Island, is rerouted to Johnston Island.
* Japanese submarine *I-10* shells and sinks unarmed Panamanian-flagged motorship ***Donerail*** 200 miles SE of Hawaii. There are only eight survivors of the 33-man crew.
* Niihau islanders attempt this day and the next to transport crash-landed Japanese Zero pilot to Kauai and are frustrated by bad weather. (see 12-13 December).
* Japanese naval land attack planes bomb defense installations on Wake Island
* The remainder of the British Empire, Cuba, and Panama declare war on Japan.
* Japanese occupy Bangkok, Thailand.
* Bitter fighting for airfield at Kota Bharu in Malaya.
* River gunboat *Mindanao* (PR-8), en route from Hong Kong to Manila, captures Japanese fishing vessel and takes her 10-man Formosan crew prisoner, leaves the craft adrift, reaching her destination the following day.
* Submarine *Swordfish* (SS-193), in initial U.S. submarine attack of the war damages Japanese ship 150 miles west of Manila.

Dec 10, Wednesday.
* Cavite Navy Yard, P.I., is practically obliterated by Japanese land attack planes. Destroyers *Peary* (DD-226) and *Pillsbury* (DD-227), submarines *Seadragon* (SS-194) and **Sealion** (SS-195), minesweeper *Bittern* (AM-36), and submarine tender *Otus* (AS-20), suffer varying degrees of damage from bombs; ferry **Santa Rita** (YFB-681) is destroyed by a direct hit. Submarine rescue vessel *Pigeon* (ASR-6) tows *Seadragon* out of the burning wharf area; minesweeper *Whippoorwill* (AM-35) recovers *Peary*, enabling both warships to be repaired and returned to service. **Bittern** is gutted by fires. Antiaircraft fire from U.S. guns is ineffective. During the bombing of Manila Bay area, unarmed U.S. freighter *Sagoland* is damaged.
* While flying to safety during the raid on Cavite, a PBY is attacked by three Japanese Zero carrier fighters ; gunner Chief Boatswain Payne shoots down one, thus scoring the U.S. Navy's first verifiable air-to-air "kill" of a Japanese plane in the Pacific War.
* Japanese forces land on Camiguin Island and at Gonzaga and Aparri, Luzon. Off Vigan, minesweeper *W.10* is bombed and sunk by a USAAF P-35; destroyer *Murasame* and transport *Oigawa Maru* are strafed, the latter set afire. USAAF B-17s bomb and damage light cruiser *Naka* and transport *Takao Maru*; the latter is run aground. Off Aparri, minesweeper *W.19* is bombed and light cruiser *Natori* is damaged.
* British battleship HMS **Prince of Wales** and battlecruiser HMS **Repulse** (VAdm Sir Tom Phillips, RN) are sunk by Japanese land-based Nell and Betty bombers off Kuantan, Malaya. Four U.S. destroyers that had been sent to help screen Phillips's ships arrive at Singapore too late to sortie with the British force; they search unsuccessfully for survivors before returning to Singapore.
* Governor of Guam (Capt. McMillin) surrenders the island to Japanese invasion force. District patrol craft *YP-16* and *YP-17*, 13 open lighters, dredge *YM-13, water barges,* and auxiliary *Robert L. Barnes* (AG-27) are all lost to the Japanese occupation of this American Pacific possession.
* Scout bomber from carrier *Enterprise* (CV-6) sinks Japanese submarine *I-70* in Hawaii area.
* Japanese naval land attack planes bomb Marine installations on Wake Island. During the attack, a USMC pilot shoots down a Mitsubishi G3M2 Type 96 land attack plane (Nell); this is the first USMC air-to-air "kill" of the Pacific War. Japanese submarines *RO-65, RO-66,* and *RO-67* arrive off Wake. Shortly before midnight, submarine *Triton* (SS-201), patrolling south of the atoll, encounters a Japanese scout warship.

Dec 11, Thu.
- Germany and Italy declare war on the United States.
- United States declares war on Germany and Italy.
- Secretary of the Navy Knox arrives on Oahu to personally assess the damage inflicted by the Japanese on 7 December.
- Submarine *Triton* (SS-201), patrolling south of Wake Island, unsuccessfully attacks a Japanese ship.
- Wake Island garrison repulses Japanese invasion force; Marine shore battery gunfire sinks destroyer, **Hayate**, and damages three other ships ; USMC F4F Wildcats bomb and sink destroyer **Kisaragi** and strafe and damage light cruiser *Tenryu* and armed merchant cruiser *Kongo Maru*. Later, USMC F4F bombs and most likely damages submarine *RO-66* south of Wake. U.S. submarines deployed off Wake, *Triton* to the south and *Tambor* (SS-198) to the north, take no active part in the battle. Following the failed assault, Japanese naval land attack planes bomb Marine gun batteries.
- Japanese submarine *I-9* shells U.S. freighter **Lahaina** about 800 miles NE of Honolulu. All American peacetime shipping is unarmed.
- Heavy bombing of Penang, Philippines, Malaya, and Hong Kong.
- Japanese make landings at Legaspi, Luzon.

Dec 12, Fri.
- Secretary of the Navy Knox departs Oahu after inspecting the damage done by the Japanese attack of 7 December.
- Downed Japanese naval pilot Nishikaichi Shigenori, armed by Harada Yoshio, a Japanese resident of Niihau, begins to terrorize the inhabitants of the island into returning papers taken on 7 December. The islanders flee to hills.
- Japanese reconnaissance flying-boats bomb Wake Island in pre-dawn raid. Later in the day, land attack planes bomb Wake.
- Heavy fighting in Philippines and Malaya.

Dec 13, Sat.
- *Indianapolis* force (VAdm Wilson Brown) returns to Pearl Harbor after searching for the Japanese fleet.
- Congress, to meet the demand for trained enlisted men, authorizes the retention of enlisted men in the Navy upon the expiration of their enlistments when not voluntarily extended.
- Japanese naval land planes attack Subic Bay area and airfields in Philippines and bomb shipping in Manila Bay.
- Occupation of Niihau, Hawaii, by downed Japanese Zero pilot Nishikaichi Shigenori ends. Nishikaichi burns his plane (it will not be until July 1942 that the U.S. Navy will be able to obtain an intact Zero to study) and the house in which he believes his confiscated papers are hidden. Later, in a scuffle, local Hawaiian, Benny Kanahele, is shot three times ; he picks up Nishikaichi bodily and dashes the pilot's head into a stone wall, killing him. Harada Yoshio, the

Japanese resident of Niihau who had allied himself with the pilot, commits suicide. Kanahele survives his injuries.
- Japanese advance in NW Malaya. Chinese attack the troops at Hong Kong.

Dec 14, Sun.
- TF 11 (VAdm Wilson Brown, Jr.), comprising carrier *Lexington* (CV-2), three heavy cruisers, nine destroyers, and oiler *Neosho* (AO-23), sails for the Marshall Islands with orders to create a diversion to cover TF 14's attempt to relieve Wake Island.
- Japanese reconnaissance flying boats bomb Wake Island. Later in the day, naval land attack planes raid Wake, bombing airfield installations.

Dec 15, Mon.
- Seaplane tender *Tangier* (AV-8), oiler *Neches* (AO-5), and four destroyers sail from Pearl Harbor for Wake Island.
- *Saratoga* arrives at Pearl Harbor.
- Japanese reconnaissance flying boats bomb Wake Island.
- Johnston Island is shelled by Japanese submarine *I-22*.
- Kahului, Maui, Territory of Hawaii, is shelled by Japanese submarine.
- British pushed back from N.W Malaya, Singapore, and Hong Kong.

Dec 16, Tue.
- Carrier *Yorktown* (CV-5) and escorts depart Norfolk, Virginia, to return to and reinforce the Pacific fleet.
- TF 14 (RAdm Frank Jack Fletcher), comprising heavy cruisers *Astoria* (CA-34) (flagship), *Minneapolis* (CA-36), and *San Francisco* (CA-38); carrier *Saratoga* (CV-3), and four destroyers sails from Pearl Harbor. These ships will overtake the force formed around *Tangier* (AV-8) and *Neches* (AO-5) and their consorts to relieve Wake Island.
- Japanese Pearl Harbor Attack Force detaches carriers *Hiryu* and *Soryu*, heavy cruisers *Tone* and *Chikuma*, and two destroyers to reinforce a planned second attack on Wake Island.
- Japanese naval land attack planes bomb Wake.
- Submarine *Tambor* (SS-198), damaged, retires from the waters off Wake.
- *Enterprise* task force returns to Pearl Harbor after searching for Japanese.

Kaswanishi H6K "Mavis" flying boat reconnaissance bomber

Dec 17, Wed.
- VAdm William S. Pye, Commander, Battle Force, becomes acting Commander in Chief Pacific Fleet, pending the arrival of RAdm Chester W. Nimitz, who is ordered on this date to relieve Adm Husband E. Kimmel.
- Seaplane from Japanese submarine *I-7* reconnoiters Pearl Harbor.
- USMC SB2U Vindicators, led by a plane-guarding PBY Catalina (no ships are available to plane-guard the flight), arrive at Midway, completing the longest over-water massed flight (1,137 miles) by single-engine aircraft. The squadron had been embarked in *Lexington* (CV-2) when the outbreak of war canceled that projected ferry mission on 7 Dec 1941.
- *Lexington* ordered north to join with *Saratoga* on Wake relief mission.
- Japanese submarine **RO-66** is sunk in collision with sistership *RO-62* off Wake Island.
- Three Japanese landings in Sarawak, Borneo ; British withdraw, destroy refineries.

Dec 18, Thu.
- Congress passes first War Powers Act granting the government more power.
- Dutch Dornier-24 flying boat bombs and sinks Japanese destroyer, ***Shinonome***, off Miri, Borneo.
- Dutch and Australian troops land in Portuguese Timor to forestall Japanese.
- Seaplane tender *Wright* transports 126 marines with their gear to Midway.

Dec 19, Fri.
- Japanese naval land attack planes bomb Wake Island.
- Japanese take Penang, Malaya.
- Fighting in the Philippines.

Dec 20, Sat.
- TF 8 (VAdm William F. Halsey, Jr.), formed around carrier *Enterprise* (CV-6), heavy cruisers, and destroyers, sails from Pearl Harbor proceeding to waters west of Johnston Island and south of Midway to cover TF 11 and TF 14 operations.
- Japanese troops land at Davao, Mindanao, P.I.
- U.S. tanker ***Emidio*** is shelled, torpedoed and sunk by Japanese submarine *I-17* about 25 miles west of Cape Mendocino, California.
- U.S. tanker *Agwiworld* is shelled by Japanese submarine *I-23* off the coast of California.
- Adm Ernest King, Commander, Atlantic Fleet, to Commander, U.S. Fleets.
- Damaged battleships *Pennsylvania, Maryland*, and *Tennessee* depart Pearl for West Coast shipyards.

Dec 21, Sun.
- Planes from carriers *Soryu* and *Hiryu* bomb Wake Island. Later that day, land attack planes bomb Wake Island.

Dec 22, Mon

* President Roosevelt and British Prime Minister Churchill open discussions in Washington (Arcadia Conference) which lasts into January 1942, results in a formal American commitment to the "Germany First" strategy.
* Japanese attack planes, covered by fighters, from carriers *Soryu* and *Hiryu*, bomb Wake Island for the second time. The last two flyable USMC F4F Wildcats intercept the raid. One F4F is shot down, the other is badly damaged.
* American troops arrive at Brisbane in convoy escorted by heavy cruiser *Pensacola* (CA-24). This is the first U.S. Army troop detachment to arrive in Australia. They had been en transit to the Philippines.
* Japanese submarine *I-19* shells the U.S. tanker *H. M. Storey* southwest of Cape Mendocino, California. The American ship escapes.
* Japanese commence invasion of Luzon, landing troops at Lingayen, Philippine Islands. Fierce fighting.

Dec 23, Tue.

* Wake Island is captured by naval landing force that overcomes gallant resistance offered by the garrison. Planes from carriers *Hiryu* and *Soryu*, as well as seaplane carrier *Kiyokawa Maru* provide close air support for the invasion.
* Uncertainty over the positions of and number of Japanese carriers and reports that indicate Japanese troops have already landed, compel VAdm William Pye, Acting Commander in Chief Pacific Fleet, to recall *Saratoga* (CV-3, TF 14, RAdm Frank Fletcher) while it is 425 miles from Wake Island.
* Palmyra Island is shelled by Japanese submarines *I-71* and *I-72*.
* U.S. tanker **Montebello** is torpedoed and sunk by Japanese submarine *I-21* about four miles south of Piedras Blancas light, California. *I-21* machine-guns the lifeboats but miraculously inflicts no casualties. *I-21* later also shells U.S. tanker *Idaho* near the same location.
* Japanese submarine *I-17* shells U.S. tanker *Larry Doheny* southwest of Cape Mendocino, California.
* USAAF B-17s bomb Japanese ships in Lingayen Gulf and off Davao. USAAF P-40s and P-35s strafe landing forces in San Miguel Bay, Luzon, damaging destroyer *Nagatsuki*.
* Japanese troops land at Kuching, Sarawak, Borneo.
* Anglo-U.S. War Council held with P.M. Churchill at the White House.

Dec 24, Wed.
• U.S. freighter *Absaroka* is shelled by Japanese submarine *I-19* about 26 miles off San Pedro, California. Steamship *Dorothy Philips* is shelled by submarine *I-23* off Monterey Bay, CA. Each is beached and recovered..
• Seaplane tender *Wright* (AV-1) disembarks Marine reinforcements at Midway.
• Second Marine Brigade is formed in California to defend American Samoa.
• Japanese land at Lamon Bay, Luzon, and thrust towards Manila.

Dec 25, Thu.
• U.S. Asiatic Fleet headquarters moves from Manila to Java.
• British surrender Hong Kong.
• Japanese land at Jolo, P.I.
• Carrier *Saratoga* (CV-3) diverted from the attempt to relieve Wake Island, flies off USMC F2As Buffaloes to Midway. These are the first fighter aircraft based there.

Dec 26, Fri.
• Manila is declared an open city but Japanese bombing continues unabated.
• Seaplane tender *Tangier* (AV-8), diverted from the attempt to relieve Wake Island, disembarks a defense battery and the ground echelon of the Marine air squadron at Midway.
• Admiral Thomas Hart leaves Manila for Soerabaja, Java, in the submarine USS *Shark* (SS-174).
• Local Philippine naval defense forces move to Corregidor.

Dec 27, Sat.
• *Lexington* and *Saratoga* return to Pearl Harbor from the attempted relief of Wake Island, their orders having been canceled when the island fell.
• U.S. tanker *Connecticut* is shelled by Japanese submarine *I-25* about 10 miles west of the mouth of the Columbia River.
• Kuala Lumpur, Malaya, is raided by Japanese bombers.

Dec 28, Sun.
• Destroyer *Peary* (DD-226) is damaged when mistakenly bombed and strafed by RAAF Hudsons off Kina, Celebes, N.E.I.
• Recruiting begins for Naval Construction Battalions, "Seabees".

Dec 29, Mon.
• Corregidor is bombed for the first time by Japanese naval land attack planes ending "normal" above-ground living there.
• Great battle raging in north Hunan, China.
• *Pennsylvania* (BB-38) arrives San Francisco for repairs.

196 *Fletcher, Task Force Commander*

Dec 30, Tue.
- Japanese submarine *I-1* shells Hilo, Hawaii.
- *Thresher* (SS-200) departs Pearl headed for the Marshalls and Marianas.
- *Tennessee* (BB-43) and *Maryland* (BB-46) enter Puget Sound Yard for repair.
- *Yorktown* (CV-5) reaches San Diego, CA; she had departed Norfolk, VA, on 16 December.

Dec 31, Wed.
- Adm Chester Nimitz assumes command of Pacific Fleet in ceremonies on board submarine *Grayling* (SS-209) at Pearl Harbor.
- Japanese submarines shell the islands of Kauai, Maui, and Hawaii.
- *Saratoga* (CV-3) and Task Force 11 depart Pearl Harbor on patrol. Japanese submarine *I-6* will torpedo the carrier on 11 January, forcing her retirement to Pearl and repair at Bremerton for four months.
- U.S. cargo/passenger ship *Ruth Alexander*, en route from Manila to Balikpapan, Borneo, is bombed and irreparably damaged by Japanese flying boat in Makassar Strait, N.E.I.
- Battles rage around Manila. The Japanese will occupy the city within days.
- *Yorktown* (CV-5) becomes flagship for RAdm Frank Fletcher's newly formed Task Force 17. The first mission will be to escort the Second Marine Brigade to American Samoa.

> Most of this section was extracted from *The Official Chronology of the U.S. Navy in Naval Historical Center.World War II*, by Robert J. Cressman, Contemporary History Branch,

NIMITZ Takes Command of Pacific Fleet

Acts of Terrorism and Atrocity by Japanese

Most of these incidents were known at the time of the relocation of enemy aliens from the coastal war zone.

Nanking, China. Over 200,000 Chinese men were used for bayonet practice, machine-gunned, or set on fire. Thousands more were murdered. 20,000 women and girls were raped, killed or mutilated. The massacre of a quarter million people was an intentional policy to force China to surrender. It did not happen. World opinion, which until this time had accepted modern Japan's desire to oversee backward China, was repelled in horror.

New officers were indoctrinated to the expectations of war by beheading Chinese captives. The last stage of the training of combat troops was to bayonet a living human and a trial of courage for the officers. Prisoners were blindfolded and tied to poles; soldiers dashed forward to bayonet their target at the shout of "Charge!"

Combat medical units moved to China where live bodies were plentiful. If the class was in sutures, a Chinaman was shot in the belly for doctors to practice. Class on amputations - then arms were removed. Living people are more instructive to work on than were cadavers; the students needed to get used to blood, screaming and movement.

Bacterial warfare experiments conducted by an infamous medical unit moved to Manchuria. Bombs of anthrax and plague were tested on Chinese cities until the results were so good that too many Japanese soldiers also died when they captured the city. This unit also practiced vivisection. Those with the stomach can look up more details of Unit 731.

Korean Comfort Women. Women and girls were forced by the Imperial Japanese Army to repeatedly provide sex for Japanese soldiers throughout Asia : are said to number between 80,000 and 200,000. Many of the victims were underage at the time, and either died in despair or suffered health impairments.

Malaya. Japanese troops decapitated 200 wounded Australians and Indians left behind when Australian troops withdrew through the jungle from Muar.

Singapore. Japanese soldiers bayonet 300 patients and staff of Alexandra military hospital 9 Feb 1942. British women had their hands behind their backs and were repeatedly raped. All Chinese residents were interviewed and 5,000 selected for execution.

Wake Island. A construction crew of 1,200 mostly Idaho youths, captured when Wake Island fell, were shipped to Japanese prison camps. Five were beheaded to encourage good behavior on the trip. The Japanese decided to keep 100 of the civilian contractors on the island to complete the airbase, which became functional in 1943. When U.S. Navy planes attacked the island, the Japanese commander executed all 96 surviving U.S. civilians.

Dutch East Indies. Those Dutch accused of resisting Japan or participating in the destruction of the oil refineries had arms or legs chopped off. 20,000 men were forced into the ocean and machine-gunned. 20,000 women and children were repeatedly raped, then many were killed.

Dutch Borneo. The entire white population of Balikpapan Is. was executed.

Java. The entire white male population of Tjepu was executed. Women were raped.

USS Edsall (DD-219) survivors are beheaded.

Philippines. Any soldier captured before the surrender was executed.
The Bataan Death March -- 7,000 surrendered men died. Those that could not keep up the pace were clubbed, stabbed, shot, beheaded or buried alive. Once the prison camp had been reached, disease, malnutrition and brutality claimed up to 400 American and Filipinos -- *each day*.

Thailand. 15,000 military prisoners and 75,000 native laborers died building a railroad between Bangkok and Rangoon. *Bridge Over the River Kwai.*

Doolittle Raid, Japan. Three of eight U.S. airmen captured were executed.

Doolittle Raid, China. Twenty-five thousand Chinese in villages through which the U.S. fliers escaped were slaughtered in a three month reign of terror.

Midway. Japanese destroyers rescued three U.S. naval aviators; after interrogation, all three were murdered. One has stuck in the head with an axe and his hand chopped as he clung to the ship's railing. Two had weighted cans tied to their legs and were thrown overboard.

Attu. Japanese troops overran the medical aid station. After killing the doctors, they bayoneted the wounded.

Makin Atoll. Nine of Carlson's Marine Raiders were left behind, hid for two weeks and surrendered. They were beheaded a few weeks later when a ship was not available to take them to a prisoner of war camp.

Milne Bay. In their few days at Milne Bay the Japanese displayed remarkable brutality. Fifty-nine local people were murdered by the Japanese, often being bayoneted while held prisoner, and in many cases being tortured. Not one of the 36 Australians captured by the Japanese survived. All were killed, and some were badly mutilated.

Bushi, *the way of the soldier*, was the creed of the Japanese in the Pacific War. It was not that long ago. The story of atrocities created under a pagan code is suppressed today in the interests of good will with a business partner. Less we forget. Civilization is only a veneer over other instincts of mankind.

One Act of Compassion.
While the Japanese were destroying the U.S. forces in the Philippines, a pilot dropped a message saying they intended to destroy the facility next to the base hospital and that we should move the patients. We did. They did.

Cleveland News -- Friday, June 5, 1942. Home Final Edition -- Page 1
President Says "Full Retaliation Will Be This Country's Policy"
Washington -- AP

"Authoritative reports are reaching this government of the use by Japanese armed forces in various localities of China of poisonous or noxious gases. I desire to make it unmistakably clear that, if Japan persists in this inhuman form of warfare against China or against any other of the United Nations, such action will be regarded by this government as though taken against the United States, and retaliation in kind and in full measure will be meted out.

"We will be prepared to enforce complete retribution. Upon Japan will rest the responsibility"
   -- Roosevelt.

History tells mass murder comes in many names, of Attila, Genghis Khan, and Tamerlane. Hundreds of Indians and settlers were slaughtered like buffalo. Within the living lifetime : Stalin purged twenty-some millions of his own people. Mao may have topped him during 1949-76. Nazis gave final solution to five or six millions. Kurds have lost millions. The Khmer Rouge killed 1.6 million during 1975-79. Unknown numbers of Muslim, intra-Muslim, and anti-Muslim in the period since 1990.

Less we forget. Hope for peace, but be prepared to resist savagery.

USN SHIP TYPES - Typical specifications early in WWII.

Desc-Type	Tons	Length	Arms	Notes
BB - Battleship	32,000	624 ft	12- 14"	
CV - Carrier, Fleet	15,000-33,000	739-910	76 a/c 91 a/c	Flight deck length
CVE- Carrier, Escort	23,000	553	34 a/c	4 tanker conversions
CVE- Carrier, Escort	7,800	495	21 a/c	10 merchant conversion
CA - Heavy Cruiser	9,950	588	9- 8"	
CL - Light Cruiser	9,700	600	15- 6"	
CLaa -Anti-Air Cruiser	6,000	541	16- 5"	
DD - Destroyer	1,620	348	4- 5"	
SS - Submarine	1,475	299	1- 3"	10 torpedo tubes
PG - Patrol Gunboat	1,270	240	4- 6"	Corvettes
PR - River Gunboat	450	200	2- 3"	China and Philippines
SC - Sub Chaser	90	110	1- 3"	Harbor defense
PT - Motor Torpedo Boat	38	80	1-20mm, 2- .50	2-4 torpedoes

Auxiliary Ships

Many warship types and most auxiliary types have three letters indicating a specialized version of the basic type.

xxA- Attack or armed version of an auxiliary type
    AP - Auxiliary, Personnel -- troop transport
    APA - Auxiliary, Personnel, Attack (invasion troopship). *"Away all boats"*

AD - Destroyer tender	AO - Oiler	LCx - Landing Craft
AE - Ammo, explosives transport	AP - Personnel, troop transport	LSx - Landing Ship
AF - Provisions ship	AR - Repair ships	LVx - Landing Vehicle
AG - Miscellaneous auxiliary	AS – Submarine Tenders	WPx – Coast Guard Cutter
AH - Hospital ship	AT - Tugboats	
AK - Cargo ship	AV - Seaplane Tender	
AM - Minesweepers	"V" is Navy speak for aircraft	

Reference 201

An individual ship or group of ships can be reclassified when converted for special duty. Many destroyer and escort types become a different type with a "D" suffix and with "fast" added to the type name. Example: An AP is the troopship designation, but when actually a converted destroyer, it becomes APD, a fast troop transport as used for the landing of raiding parties.

Designations change from time to time. Early WWII designation and typical specifications are used here. Ships tended to get bigger and better armed as the war progressed. Many changes took place after the war.

USS Wakefield AP-21 – troopship (with lifeboats)

USS Leonard Wood APA 12 --Attack Transport (w/landing craft)

How U.S. Navy Ships Were Named in WWII

The Navy Department had standard rules for naming Navy ships during WWII. There are some exceptions to these rules, but in general they are closely followed. No vessel was named in honor of a living person.

Battleships (BB) --named for states by act of Congress. (*Nevada, Iowa.*)

Cruisers (CA, CL) --named for cities and towns in the United States and capitals of United States possessions and territories. (*Atlanta, Brooklyn, Chicago, Salem, Juneau*)

Battlecruisers (CB) --named for territories and insular possessions of the United States. (*Alaska, Guam*)

Aircraft Carriers (CV, CVL) --usually named for famous ships formerly on the Navy list; and important battles. (*Ranger, Bonhomme Richard, Lexington, Saratoga*)

Aircraft Carriers (CVE) --named for bays, islands, and sounds of United States and for battles of World War II. (*Tulagi, St. Joseph Bay, Guadalcanal, Block Island*)

Destroyers (DD) --named for deceased persons in the American Navy, Marine Corps, and Coast Guard who rendered distinguished service to their country, Secretaries and Assistant Secretaries of the Navy, members of Congress who were closely identified with naval affairs, and inventors. (*Porter, Sims, Laffey, The Sullivans, Heywood, Edwards, Edison*)

Submarines (SS) --named for fish and other denizens of the deep. (*Barracuda, Drum, Perch, Silversides, Sturgeon, Whale*)

Destroyer Escorts (DE) --named in honor of personnel of Navy, Marine Corps, and Coast Guard killed in enemy action in World War II. (*Lloyd E. Acree, Douglas A. Munro, Parle, Slater, Stewart, Traw*)

Frigates (PF) --named for cities and towns in the United States, and towns of United States possessions and territories. (*Newport, Alexandria, Annapolis*)

Gunboats (PG) --named for small cities; logical and euphonious words. (*Asheville, Erie, Fury, Action*)

Yachts; Yachts, coastal (PY, PYc) -- named for old ships formerly in the Navy; gems; also logical and euphonious words. (*Siren, Emerald, Valiant, Sturdy*)

Auxiliary Ships

Destroyer Tenders (AD)--named for localities and areas of the United States. (*Prairie, Dixie, Everglades*)

Ammunition Ships (AE) --words which are suggestive of fire and explosives; and names of volcanoes. (*Pyro, Nitro, Lassen, Vesuvius*)

Hospital Ships (AH) --logical and euphonious words. (*Mercy, Relief, Samaritan*)

Cargo Ships (AK) --named for astronomical bodies (*Rugulas, Sirius, Capella*) and counties of the United States. (*Arlington*)

Minesweepers (AM) --named for birds; logical and euphonious words. (*Raven, Cardinal, Pursuit, Adroit, Hazard*)

Oilers (AO) --named for Indian names of rivers. (*Platte, Rapidan, Brazos*)

Naval Transports (AP, APA) --named in honor of deceased Marine Corps officers, counties in the United States; places of historical interest; signers of the Declaration of Independence; famous women of history; and famous men of foreign birth who rendered aid to our country in her early struggle for independence. *Le Jeune, Middleton, Elizabeth Stanton, Doyen, Freemont*)

Repair Ships (AR) --named for characters in mythology. (*Medusa, Vestal, Rigel*)

Submarine Tenders (AS) --named in honor of pioneers in submarine development and also for sea deities in mythology. (*Fulton, Holland, Pelias*)

Submarine Rescue Vessels (ASR) --named for birds. (*Falcon, Greenlet*)

Seaplane Tenders (AV) --named for sounds. (*Tangier, Puget Sound*)

Seaplane Tenders (AVP) --named for bays, straits, inlets. (*Barnegat, Greenwich Bay, Biscayne, Bering Strait, Cook Inlet*)

Net Tenders (YN) --named for trees, and (Tug Class) India chiefs and other noted Indians. (*Hackberry, Holly, Ebony, Yaupon, Tesota*)

Ocean-going Tugs (ATF) --named for Indian tribes. (*Cherokee, Navajo, Apache, Choctaw*)

Harbor Tugs (large) (YTB) --named for Indian chiefs and words of Indian dialect. (*Tecumseh, Pawtucket*)

Imperial Japanese Navy Ship Names

Heavy Cruiser IJN **Mogami** – Mogami River.

Battleships – Ancient Provinces
Carriers – Dragons and Birds
Heavy Cruisers – Mountains
Light Cruisers – Rivers
Destroyers (1st class) - Meteorological names
Destroyers (2nd class) – Trees, Flowers, and Fruits
Torpedo Boats – Birds
Minelayers – Islands, Straits, Channels (old layers were birds)

Destroyer IJN **Yunagi** -- "Evening Calm"

U.S. Battleships of World War II

ID	Name	Commission	Tons	Guns	on 7Dec41
BB-33	Arkansas	17Sept1912	27,243	6x2 12"	Maine
BB-34	New York	15Apr1915	27,000	5x2 14"	Newfoundland
BB-35	Texas	12Mar1914	27,000	5x2 14"	Maine
BB-36	Nevada	11Mar1916	27,500	2x3,2 14"	**Beached P.H.**
BB-37	*Oklahoma*	2May1916	27,500	2x3,2 14"	*Pearl Harbor*
BB-38	**Pennsylvania**	**12June1916**	**31,400**	**4x3 14"**	damaged
BB-39	*Arizona*	17Oct1916	31,400	4x3 14"	*Pearl Harbor*
BB-40	**New Mexico**	**20May1918**	**32,000**	**12 14"**	neutrality patrol
BB-41	**Mississippi**	**18Dec1917**	**32,000**	**12 14"**	**Iceland**
BB-42	**Idaho**	**24Mar1919**	**32,000**	**12 14"**	**Iceland**
BB-43	**Tennessee**	**3June1920**	**33,190**	**12 14"**	damaged
BB-44	California	10Aug1921	32,300	12 14"	***Sunk***, refloated
BB-45	**Colorado**	**30Aug1923**	**32,600**	**8 16"**	W.Coast overhaul
BB-46	**Maryland**	**21July1921**	**32,600**	**8 16"**	damaged
BB-47	Washington	uncompleted	32,600	8 16"	*Treaty'23*
BB-48	**West Virginia**	**1Dec1923**	**33,590**	**8 16"**	***Sunk***, refloated
BB-49	South Dakota	uncompleted	43,200	12 16"	*Treaty'23*
BB-50	Indiana	uncompleted	43,200	12 16"	*Treaty'23*
BB-51	Montana	uncompleted	43,200	12 16"	*Treaty'23*
BB-52	North Carolina	uncompleted	43,200	12 16"	*Treaty'23*
BB-53	Iowa	uncompleted	43,200	12 16"	*Treaty'23*
BB-54	Massachusetts	uncompleted	43,200	12 16"	*Treaty'23*
BB-55	**North Carolina**	**9Apr1941**	**35,000**	**9 16"**	Atlantic
BB-56	**Washington**	**15May1941**	**35,000**	**9 16"**	Atlantic
BB-57	**South Dakota**	**20Mar1942**	**35,000**	**9 16"**	pre-comm
BB-58	**Indiana**	**30Apr1942**	**35,000**	**9 16"**	pre-comm
BB-59	Massachusetts	12May1942	35,000	9 16"	pre-comm
BB-60	Alabama	16Aug1942	35,000	9 16"	pre-comm

BB-61	Iowa	22Feb1943	45,000	9	16"	Jan'44
BB-62	New Jersey	23May1943	45,000	9	16"	Jan'44
BB-63	Missouri	11June1944	45,000	9	16"	Dec'44
BB-64	Wisconsin	16April1944	45,000	9	16"	Oct'44
BB-65	Illinois	canceled	45,000	9	16"	1/4 complete
BB-66	Kentucky	canceled	45,000	9	16"	2/3 complete
BB-67	Montana	canceled	65,000	12	16"	
BB-68	Ohio	canceled	65,000	12	16"	
BB-69	Maine	canceled	65,000	12	16"	
BB-70	New Hampshire	canceled	65,000	12	16"	
BB-71	Louisiana	canceled	65,000	12	16"	
CB-1	Alaska	17Jun44	27,500	9	12"	Dec'44
CB-2	Guam	17Sep44	27,500	9	12"	Feb'45
CB-3	Hawaii	suspended	27,500	9	12"	4/5 complete

Bold means operated in the Pacific during the first year.
Italic means lost in the war.
Montana Class of five battleships was canceled July'43.
CB type officially large cruiser, commonly battlecruiser.

There were no operational battleships in the Pacific after Pearl Harbor. *Mississippi* and *Idaho* departed Iceland two days later by way of Norfolk for San Francisco, arriving 22/31Jan42, six weeks later. *New Mexico* also returned from the Atlantic and the three formed Task Force 1 based out of San Francisco to patrol and escort convoys to Hawaii. *Tennessee* and *Maryland* were repaired from Pearl Harbor and sailed from Puget Sound Navy Yard 26Feb42. *Colorado* finished overhaul a month later, at about the same time as *Pennsylvania* finished repair. The new construction *North Carolina* (BB-55) arrived in June restoring the battle fleet and to its prewar number. The new construction *Washington* (BB-56) and *South Dakota* (BB-57) arrived in August to fight at Guadalcanal and *Indiana* (BB-58) arrived in November.

The newest battleship, *Washington*, and newest aircraft carrier, *Wasp* (CV-7), were involved in the prewar Atlantic fiction of "neutrality patrol" in the Atlantic. Both departed to reinforce the British Home Fleet as soon as they could, in March 1942, while new carrier *Hornet* (CV-8) departed for the Pacific about the same time and carried Doolittle's Raiders for Tokyo. *Washington* performed convoy escort to northern Russia and North Atlantic. *Wasp* made two ferry runs through Gibraltar to Malta. When *Lexington* (CV-2) was lost at the Coral Sea, *Wasp* was hurried home to replace her. *Wasp* and new battleship *North Carolina* departed Norfolk while the Battle of Midway was going on, where *Yorktown* (CV-5) was lost in June. New *South Dakota* was ready to depart for the Pacific in mid-August, followed a week later by *Washington* which had been recalled when the leadership became convinced that the Pacific War deserved more of the new equipment.

The Navy in the South Pacific was down to one active battleship for a few weeks, twice, late in 1942. *North Carolina* had been torpedoed 16Sept and was in repair at Pearl Harbor. *South Dakota* was damaged in the inconclusive Battle of Santa Cruz, 26 Oct, and under repair at New Caldonia. Only *Washington* was active in the South Pacific. The older *Colorado* (BB-45) was rushed to Fiji by 8Nov. *Washington* and mostly repaired *South Dakota* fought on the second night of the Naval Battle of Guadalcanal, 13-14 Nov, where *South Dakota* was badly damaged and returned to NY. *Washington* was again the lone, fast battleship until 28Nov when *Indiana* (BB-58) arrived fresh from shakedown. Old battleships *Arkansas* (BB-33), *New York* (BB-34), and *Texas* (BB-35) remained in the Atlantic and were joined by *Nevada* (BB-36) in mid-1943, where their slower speed was not a handicap for troop convoy escort service.

New escort carriers were mostly assigned to the Atlantic while the large tanker-conversion carriers saw Atlantic service through the invasion of North Africa in October then joined the Pacific fleet in early 1943.

Planned *Illinois* (BB-65) was intended as first of the larger Montana class battleship but was constructed as an Iowa class in a rush to provide protection to the new *Essex* class carriers scheduled to start commissioning in 1943. *Essex* (CV-9) did enter operations at Marcus Island in Aug'43 and attacked Rabaul in November without battleship escort.

History.

After WWI, super-dreadnoughts were introduced from the lessons learned in the Battle of Jutland, 1 June 1916. After that war an international moratorium on new battleships went into effect. As part of the treaty the U.S. scrapped all pre-dreadnoughts and seven dreadnought battleships, leaving five small and seven larger dreadnought types plus three super-dreadnought types which were under construction. By the end of the treaty period, the U.S. was three battleships below treaty allowance. The treaty expired in 1936 with Japan immediately starting to build some giant capital ships, double the Treaty limit sizes. The U.S. followed by starting on six new battleships, though smaller because they had been designed to treaty limitations. Eight of the existing sixteen, including *Maryland* and *West Virginia*, were sunk or damaged at Pearl Harbor. A fear that the new ships would strengthen the U.S. Pacific Fleet encouraged Japan to strike as her new, large ships came into service while she still held a large edge in the Pacific in both battleships and aircraft carriers.

Naval Battleship Monuments

Name	Number	Called	Location	Battles
Utah	BB-31		under Pearl Harbor, HI	Vera Cruz, WWI.
Texas	BB-35	"Tex"	San Jacinto, TX	Vera Cruz, WWI, WWII
Arizona	BB-39		Pearl Harbor, HI	
North Carolina	BB-55	Showboat	Wilmington, NC	WWII
Massachusetts	BB-59	Big Mamie	Fall River, MA	WWII
Alabama	BB-60	Mighty A	Mobile, AL	WWII
Iowa	BB-61	Big Stick	Los Angles, CA	WWII, Korea *
New Jersey	BB-62	Big J	Camden, NJ	WWII, Korea, Vietnam, Lebanon
Missouri	BB-63	Mighty Mo	Pearl Harbor, HI	WWII, Korea, Desert Storm
Wisconsin	BB-64	Whiskey	Norfolk, VA	WWII, Korea, Desert Storm *

* Iowa and Wisconsin were maintained in case needed as robust missile ships.

USS Alabama BB-60 is at Battleship Memorial Park, Mobile, Alabama

U.S. Aircraft Carriers of World War II

Fleet Carriers

	Name CoCo	Commision	Tons	Feet	On 7Dec41	Lost
CV -1 \AV-3	Langley	1922, 1936 converted	19,360	542	Philippines.	Java
CV -2	Lexington	14Dec27	33,000	910	to Midway	Coral Sea
CV -3	Saratoga	16Nov27	33,000	910	to San Diego	7 ✯
CV -4	Ranger	04Jun34	14,500	769	to Norfolk	2 ✯
CV -5	Yorktown	30Sep37	19,800	827	Norfolk.	Midway
CV -6	Enterprise	12May38	19,800	827	to Pearl Harbor	20 ✯
CV -7	Wasp	25Apr40	14,700	739	Bermuda.	Guadalcanal
CV -8	Hornet	20Oct41	20,000	809	Norfolk.	Santa Cruz
CV -9	Essex	31Dec42	27,100	874	13✯	
CV-10	Yorktown II	15Apr43	27,100	874	11✯	
CV-11	Intrepid	16Aug43	27,100	874	5+✯ + Vietnam	
CV-12	Hornet II	29Nov43	27,100	874	7 ✯	
CV-13	Franklin	31Jan44	27,100	874	4 ✯ + 4 Korea	
CV-14	Ticonderoga	08May44	27,100	885	5 ✯ + 12 Vietnam	
CV-15	Randolph	09Oct44	27,100	885	3 ✯	
CV-16	Lexington II	17Feb43	27,100	874	11✯	
CV-17	Bunker Hill	24May43	27,100	874	11✯	
CV-18	Wasp II	24Nov43	27,100	874	8 ✯	
CV-19	Hancock	15Apr44	27,100	885	4 ✯	
CV-20	Bennington	06Aug44	27,100	874	3+✯	
CVL-22	Independence	14Jan43	11,000	619	8 ✯ (cruiser hulls)	
CVL-23	Princeton	25Feb43	11,000	619	9 ✯	Leyte
CVL-24	Belleau Wood	31Mar43	11,000	622	12✯ France 1953-'60	
CVL-25	Cowpens	28May43	11,000	622	12✯	
CVL-26	Monterey	17Jun43	11,000	622	11✯	
CVL-27	Langley II	31Aug43	11,000	619	8 ✯ France 1951-'63	
CVL-29	Bataan	17Nov43	11,000	619	6 ✯ + 7 Korea	
CVL-30	San Jacinto	15Dec43	11,000	622	5 ✯	

CV-31	BonHomme Richard	26Nov44	27,100	874	1 ✯ + 5 Korea
CV-38	Shangri-La	15Sep44	27,100	885	2 ✯ + 3 Korea
CV	"Robin"	May43 to	Jul43	Loan	HMS *Victorious*

-------- Commissioned Too Late to Participate in the War --------

CV-21	Boxer	16 Apr 45	27,100	885	Korea
CV-32	Leyte	11 Apr 46	27,100	888	Korea
CV-33	Kearsarge	02 May46	27,100	888	Korea
CV-34	Oriskany	25 Sep 50	27,100	888	Korea
CV-35	Reprisal	Not completed			
CV-36	Antietam	28 Jan 45	27,100	885	Korea
CV-37	Princeton	18 Nov 45	27,100	888	Korea, Vietnam
CV-39	Lake Champla	03 June 45	27,100	885	Korea
CV-40	Tarawa	08 Dec 45	27,100	885	
CVB-41	Midway	10 Sep 45	45,000	968	Vietnam, Kuwait
CVB-42	F.D.Roosevelt	27 Oct 45	45,000	968	Vietnam
CVB-43	Coral Sea	01 Oct 47	45,000	968	Vietnam, Libya
CV-45	Valley Forge	03 Nov 46	27,100	888	Korea, Vietnam
CV-47	Philippine Sea	11 May 46	27,100	885	Korea
CVL-48	Saipan	14 July 46	14,500	684	Vietnam, AGMR-1
CVL-49	Wright	09 Feb 47	14,500	684	CC-2

Canceled: CV-35, CVB-44, CV-46, CV-50 to CV-55, CVB-56, CVB-57.

✯ - Battle Stars earned in WWII.

Reference

Monuments

Name	Hull	Nickname	Location	Service
Yorktown II	CV-10	Fighting Lady	Charleston	WWII, Korea, Vietnam, Apollo
Intrepid	CV-11	Fighting "I"	New York	WWII, Vietnam, Apollo
Hornet II	CV-12	Gray Lady	Alameda Point, San Francisco	WWII, Vietnam, Apollo
Lexington II	CV-16	Blue Ghost	Corpus Christi	WWII, Cuba
Midway	CVB-41	1945-1992	San Diego	Vietnam, Desert Storm

USS **Yorktown II (CV-10)** is named for Fletcher's flagship lost at Midway.

Yorktown II is the centerpiece of Patriots Point Naval and Maritime Museum in Charlestown, South Carolina.

U.S. Cruisers of World War II

Cruisers are the intermediate sized workhorse of the Navy. Lighter than battleships and more powerful than destroyers, cruisers are thought of as able to defeat most other ships or able to run away from those they can't. Cruisers are used for scouting, antiaircraft defense, shore bombardment, battle force, and many other duties.

Heavy cruisers are defined as having a main armament of 8" guns, the maximum set by treaty. USN light cruisers have 6" guns, except a few were equipped with many 5" antiaircraft guns as their primary armament. When the treaty ended, a battlecruiser class was authorized with 12" guns.

The U.S. had 37 cruisers (18 heavy and 19 light) at the start of the war to serve all ocean fronts. Ten cruisers (7 heavy and 3 light) were sunk and 36 new (6 heavy, 22 light, and 8 antiaircraft) were commissioned in time to participate in the war. Nine light cruiser hulls were used for CVL, light aircraft carriers. About fifty more cruisers were building at the end and did not participate. The last conventional gun cruiser was *Newport News* (CA-148) commissioned 29Jan49.

Bold=Fought Pac'42 ;   *Italic*=sunk

ID	Name	Commiss	Tons	Guns	7Dec41 or date to Pacific
CL-4	Omaha	Feb 1923	7,300	12 6"	S. Atlantic
CL-5	Milwaukee	Jun 1923	7,300	12 6"	NY ; **Feb'42**
CL-6	Cincinnati	Jan 1924	7,050	10 6"	S. Atlantic
CL-7	**Raleigh**	Feb 1924	7,050	10 6"	**Pearl Harbor**
CL-8	**Detroit**	Feb 1924	7,050	10 6"	**Pearl Harbor**
CL-9	**Richmond**	July 1923	7,050	10 6"	off Chile
CL-10	**Concord**	Nov 1923	7,300	12 6"	San Diego
CL-11	**Trenton**	Apr 1924	7,300	12 6"	Balboa
CL-12	**Marblehead**	Sep 1924	7,050	10 6"	**Borneo** to Feb'42
CL-13	Memphis	Feb 1925	7,050	10 6"	South Atlantic
CP-14	Chicago (CP)	April 1889	4,500	4- 8"	1913-23 FlagSub
CA-15	Olympia (C-6)	Feb 1895	5,586	4- 8"	Shrine, Philadelphia
CL-16	-to- CL-23	1898-1905	c3,340	c6 6"	scrapped 1921,1930
CA-24	**Pensacola**	Apr 1930	9,100	10 8"	**convoy to Manila**
CA-25	**Salt Lake City**	Dec 1929	10,826	10 8"	**return Wake Island**
CA-26	*Northampton*	*May 1930*	*9,050*	*9 8"*	*return Wake Island*
CA-27	**Chester**	June 1930	9,200	9 8"	**return Wake Island**

Reference 213

CA-28	Louisville	Jan1931	9,050	9 8"	return Borneo
CA-29	*Chicago*	*Mar1931*	*9,300*	*9 8"*	*TF12 from Pearl*
CA-30	*Houston*	*July1931*	*9,050*	*9 8"*	*Panay Island, P.I*
CA-31	Augusta	Jan1931	9,050	9 8"	Newport, RI
CA-32	New Orleans	Feb1934	9,950	9 8"	Pearl Harbor
CA-33	Portland	Feb1933	9,950	9 8"	TF12 from Pearl
CA-34	*Astoria*	*Apr1934*	*9,960*	*9 8"*	*TF12 from Pearl*
CA-35	*Indianapolis*	*Nov1932*	*9,800*	*9 8"*	*Johnston Island*
CA-36	Minneapolis	May1934	9,950	9 8"	off Ohau
CA-37	Tuscaloosa	Aug1934	9,950	9 8"	Iceland; Jan'45
CA-38	San Francisco	Feb1934	9,950	9 8"	Pearl Harbor
CA-39	*Quincy*	*June1936*	*9,375*	*9 8"*	*S. Atlantic convoy*
CL-40	Brooklyn	Sep1937	9,700	15 6"	convoy escort
CL-41	Philadelphia	Sep1937	9,700	15 6"	Boston overhaul
CL-42	Savannah	Mar1938	9,475	15 6"	New York
CL-43	Nashville	June1938	9,475	15 6"	Bermuda; Mar'42
CA-44	*Vincennes*	*Feb1937*	*9,400*	*9 8"*	*S. Atlantic; Mar'42*
CA-45	Wichita	Feb1939	10,000	9 8"	Iceland. Jan'43
CL-46	Phoenix	Oct1938	10,000	15 6"	Pearl Harbor
CL-47	Boise	Aug1938	9,700	15 6"	Cebu, Philippines
CL-48	Honolulu	June1938	9,650	15 6"	Pearl Harbor
CL-49	St. Louis	May1939	10,000	15 6"	Pearl Harbor
CL-50	*Helena*	*Sep1939*	*10,000*	*15 6"*	*Pearl Harbor*
CL-51a	*Atlanta*	*Dec1941*	*6,000*	*16 5"*	*Apr'42*
CL-52a	*Juneau*	*Feb1942*	*6,000*	*16 5"*	*Aug'42*
CL-53a	San Diego	Jan1942	6,000	16 5"	May'42
CL-54a	San Juan	Feb1942	6,000	16 5"	Jun'42
CL-55	Cleveland	June1942	10,000	12 6"	Jan'43
CL-56	Columbia	July1942	10,000	12 6"	Dec'42
CL-57	Montpelier	Sept1942	10,000	12 6"	Jan'43
CL-58	Denver	Oct1942	10,000	12 6"	Feb'43
CL-59/CVL-22 Independence		Jan1943	11,000	45 a/c	July'43
CL-60	Santa Fe	Nov1942	10,000	12 6"	Mar,43
CL-61/CVL-23 Princeton		*Feb1943*	*11,000*	*45 a/c*	*Aug'43*
CL-62	Birmingham	Jan1943	10,000	12 6"	Sept'43

Photo # NH 53230 USS Indianapolis at Pearl Harbor, circa 1937

Photo # NH 53562 USS Honolulu underway, February 1939

Reference

Japanese Battleships. - Ten Prewar Battleships

Name	Tons	Comm	Guns	Sunk	By	Where
Kongo	32,250	1913	4x2-14"	21Nov44	SS-315	off Formosa
Hiei	32,250	1914	4x2-14"	13Nov42	TG 67; Air	Guadalcanal
Kirishima	32,250	1915	4x2-14"	15Nov42	BB-56	Guadalcanal
Haruna	32,250	1915	4x2-14"	28July45	CV TF 38	Kure
Fuso	35,900	1915	6x2-14"	25Oct44	BB, TF 77	Surigao Sea, P.I.
Yamashiro	35,900	1917	6x2-14"	25Oct44	DD, PT	Surigao Sea, P.I.
Ise	36,500	1917/43	6x2-14"	28July45	CV TF 38	Kure
Hyuga	36,500	1917/43	6x2-14"	24July45	CV TF 38	Kure
Nagato	38,500	1920	4x2-14"	14July45	CV TF37	Yokosuka
Mutsu	38,500	1921	4x2-14"	08June43	explosion	Hiroshima
Yamato	64,000	Dec1941	3x3-18"	07Apr45	CV TF 58	Suicide Okinawa
Musashi	64,000	Aug1942	3x3-18"	24Oct44	CV TF 38	Sibuyan Sea, P.I.

IJN battleship Yamato – largest ever built

Japanese Heavy Cruisers

Name	Tons	Comm	Sunk	By	Where
Kako	8,800	1926	10Aug42	S-44	after Savo Island
Furutaka	8,800	1926	11Oct42	CA, CL	Cape Esperance
Kinugasa	9,000	1927	14Nov42	USMC SBDs	Guadalcanal
Aoba	9,000	1927	25July45	CV TF 38	Kure
Myoko	13,000	1929	25Oct44	SS-320	to Singapore
Haguro	13,000	1929	16May45	British Fleet	Malacca Strait, Malaya
Ashigara	13,000	1929	08June45	HMS Trenchant	Bangka Straits, Malaya
Atago	13,400	1932	23Oct44	SS-227	Sibuyan Sea, Leyte
Takao	13,400	1932	23Oct44 31July45	*SS-227,* Br. minisub	*Sibuyan Sea, Leyte;* Singapore
Choka	13,400	1932	25Oct44	CV, scuttled	Samar, Leyte, P.I.
Maya	13,400	1933	23Oct44	SS-247	Sibuyan Sea, Leyte
Mogami[1]	13,500	1935	25Oct44	BB, TF 77	Surigao Sea, Leyte
Mikuma	12,500	1936	06June42	CV TF 16	Midway
Suzuya	12,500	1937	25Oct44	CV, scuttled	Samar, Leyte, P.I.
Kumano	13,500	1937	25Nov44	SS, CV-14	Philippines
Nachi	13,000	1938	05Nov44	CV TF 38	Philippines
Tone[2]	11,200	1938	25July45	CV TF 38	Kure
Chikuma[2]	11,200	1939	25Oct44	CV, scuttled	Samar, Leyte, P.I.

-----

[1] *Mogami* was seriously damaged at Midway, escaped to return as a seaplane cruiser with 6- 8" guns forward and catapults aft with room for seven seaplanes.

[2] *Tone* and *Chikuma* were seaplane cruisers. They had a full complement of 8- 8" guns crowded forward and catapults for seaplanes aft; and six seaplanes. The two ships always operated together as scouts for the carrier fleet.

Main armament of the 12 built in the 1930s was 10 to 15 8" guns, 12 torpedo tubes, with 1 to 3 floatplanes

Furutaka = Hurutaka ; Tikuma = Chikuma ; Ahigara = Asigara ;
Nachi = Nati ; Choka = Takao

Japanese Light Cruisers

Name	Tons	Comm	Sunk	By	Where
Tenryu	3,300	1919	18Dec42	SS-218	Off Madang Harbor, NG
Tatsuta	3,300	1919	13Mar44	SS-381	E of Honshu
Oi	5,700	1920	19July44	SS-249	South China Sea
Kuma	5,700	1920	11Jan44	HMS Tallyho	Penang, Malaya
Tama	5,700	1921	25Oct44	CV,SS-368	Leyte
Kiso	5,700	1921	13Nov44	CV TF 38	Manila
Kitakami	5.760	1921	24July45	CV TF 38	Kure
Nagara	5,760	1922	07Aug44	SS-246	S of Nagasaki, Japan
Natori	5,760	1922	18Aug44	SS-365	E of Samar, P.I
Kinu	5,760	1922	26Oct44	CV TF 38	Leyte
Isuzu	5,760	1923	07Apr45	SS-252,SS-328	SE Celebes, N.E.I
Yura	5,760	1923	25Oct42	Multiple Air	Santa Isabel, Solomons
Yubari	3,500	1923	27Apr44	SS-242	E of Mindanao
Sendai	5,850	1924	02Nov43	TF 39	Empress Augusta Bay
Abukuma	5,760	1925	26Oct44	B-24s	Leyte
Jintsu	5,850	1925	13July43	TG-36.2	Kolombangara, Sol
Naka	5,850	1925	17Feb44	TF 58	W of Truk
Ioshima	2,500	1932	19Sep44	SS-235	E of Honshu
Yasojima	2,500	1936	25Nov44	CV TF 38	W of Luzon
Katori[1]	6,000	1940	17Feb44	TF 58	NE of Truk
Kashima	6,000	1940	survived		
Kashii	6,000	1941	12Jan45	CV TF 38	SE IndoChina
Agano	6,650	1942	16Feb44	SS-305	N of Truk
Yahagi	6,650	1943	07Apr45	CV TF 58	East China Sea
Nohiro	6,650	1943	26Oct44	CV TF 38	Leyte
Oyodo[2]	8,200	1943	28July45	CV TF 38	Kure
Sakawa	6,650	1944	survived		

[1] *Katori, Kashima,* and *Kashii* were built as training or flag ships, used as A/S escorts.
[2] *Oyodo* was a seaplane light cruiser with 6- 6" guns forward and catapults aft.

Japanese Light Cruiser YURA

5,170 tons, 7- 5.5" guns, 8- 21" tubes; 1 floatplane.

Yura, flagship of DesRon 4, led a bombardment force of destroyers from Shortland to Guadalcanal on 25 October 1942. Off Santa Isabel Island in the Solomons, she was attacked by five SBD Dauntless dive bombers from Guadalcanal and hit aft by two bombs near the engine room. She flooded and settled by the stern. After receiving reports of the attack, Vice Admiral Mikawa canceled Rear Admiral Takama's bombardment mission. The wounded ship was then attacked by a succession of Navy, Marine, and Army Air Force planes. After her crew was taken off, destroyers *Harusame* and *Yudachi* scuttle *Yura* with torpedoes. She broke in two and the forward portion sank; the stern portion was sunk off Savo Island by gunfire.

Light Cruiser AGANO

6.650 tons; 35 knots; 6- 6", 8- 24" tubes; 2 floatplanes.

Most Japanese light cruisers were command ships of destroyer or submarine squadrons. *Agano* fought in New Guinea and Guadalcanal. Torpedoed near Truk, her crew was lost when the rescue ship was also torpedoed.

Japanese Heavy Cruiser KUMANO

13,400 tons; 35 knots; 10- 7.9", 12- 21' tubes; 3 seaplanes

One of four overweight cruisers in clear violation of treaty limits.
Kumano fought in the Java Sea, Midway, Eastern Solomons, Naval Battle of Guadalcanal, Philippine Sea, and Leyte, where she lost her bow. Repaired and escorting a convoy back to the Philippines that was attacked by a U.S. wolf pack and again torpedoed. She was finally sunk by planes from Ticonderoga (CV-14) while under repair in the Philippines.

Japanese Light Carrier RYUJO

10,500 tons; 29 knots; carried 45 aircraft, operated 36.

Characteristic of Japanese ships built during the treaty period, she carried more arms than expected of a ship of her weight. Rebuilt after experiences in China. When the war started in the Pacific she was seemingly everywhere supporting the invasion of the Philippines, Malaya, Java, Burma, Indian Ocean shipping, an attack on India proper, Dutch Harbor in the Aleutians, and then to the Solomons to support a re-invasion where she attacked Henderson Field on Guadalcanal. *Ryujo* was in turn attacked by *Saratoga*. Her sinking marked the sixth carrier that Fletcher put under. The invasion fleet withdrew.

Japanese Aircraft Carriers
Ten Carriers at the time of Pearl Harbor

Name	Type (a/c)	Tons	Commis	Sunk	By	Where
Hosho	CVL (21)	7,500	27Dec22	survived-		
Akagi	CV (63)	36,000	25May27	04June42	CV	Midway
Kaga	CV (72)	36,800	21Mar28	04June42	CV	Midway
Ryujo	CVL (46)	10,500	9May33	24Aug42	CV	East Solomon
Soryu	CV (63)	17,500	29Sep37	04June42	CV	Midway
Hiryu	CV (63)	17,500	5July39	05June42	CV	Midway
Zuiho	CVL (24)	11,200	27Dec40	25Oct44	CV	Cape Engano
Shokaku	CV (72)	27,000	08Aug41	19June44	SS-244	Marianas
Taiyo	CVE (27)	16,700	15Sep41	18Dec44	SS-269	off Luzon
Zuikaku	CV (72)	27,000	25Sep41	25Oct44	CV	Cape Engano
Shoho	CVL (24)	11,200	26Jan42	07May42	CV	Coral Sea
Junyo	CV (45)	24,100	05May42	09Dec44	SS	off Sasebo
Unyo	CVE (27)	16,700	31May42	16Sep44	SS-220	S.China Sea
Hiyo	CV (45)	24,100	31July42	20June44	CV	Marianas
Chuyo	CVE (27)	16,700	25Nov42	04Dec43	SS-192	off Honshu
Ryuho	CVL (31)	13,400	28Nov42	19Mar45	CV	Kure
Hyuga	BBV (22)	conv	01Oct43	24July45	CV	Kure
Ise[1]	BBV (22)	conv	08Oct43	28July45	CV	Kure
Chiyoda	CVL (24)	11,200	31Oct43	25Oct44	CV	Cape Engano
Kaiyo	CVE (24)	15,400	23Nov43	24July45	CV	Beppu Bay
Shinyo	CVE (27)	17,500	15Dec43	17Nov44	SS-411	S.Yellow Sea
Chitose	CVL (24)	11,200	01Jan44	25Oct44	CV	Cape Engano
Taiho	CV (62)	29,300	07Mar44	19June44	SS-218	Marianas
Unryu	CV (57)	17,300	06Aug44	19Dec44	SS-395	EastChinaSea
Amagi	CV (57)	17,100	10Aug44	28June45	air	Kure
Shinano	CVB (70)	**65,000**	19Nov44	29Nov44	SS-311	Coast of Japan
Katsuragi	CV (57)	17,300	15Oct44	Moored		
Kasagi	CV	17,300	incomplete	85%		
Ibuki	CVL	14,500	incomplete	80%		
Aso	CV	17.200	incomplete	60%		
Ikoma	CV	17,200	incomplete	60%		

Reference

Hosho was Japan's first carrier and used for experimentation and training, but took part in Midway as an escort.
Kaga was converted from a battleship hull when new battleships were forbidden by Naval Treaty.

Akagi was converted from a battlecruiser hull when cruisers were limited by Naval Arms Treaty. IJN Heavy Aircraft Carrier, flagship from Pearl Harbor through Midway.
Ise and *Hyuga* were battleships with rear turrets removed after Midway and hangers and catapults installed. 8 "Paul" seaplane dive bombers and 14 "Judy" dive bombers would land on carriers. When aircrews became short, they went to sea again as battleships. Never fought as carriers.
Chiyoda and *Chitose* were converted from seaplane carriers.
Zuiho and *Shoho* were converted from seaplane tenders.
Taiyo, Unyo, and *Chuyo* were converted from passenger liners.
Shinano was a super-carrier, 65,000 tons, built on a super-battleship hull after the lessons of Midway proved the need for aircraft. She was torpedoed by *Archerfish (SS-311),* 29Nov44, and sank off Kansai traveling between her launch site and training base.

Alternate spelling and names found.

Jinyo, Shingo,	Junyo, Haytaka	Unyo, Un'yo,
Hiyo, Haytaka	Taiyo, Toiyo, Utaka, Otaka	

Pacific Commanders 1943

VAdm Fletcher -- North Pacific
Lt. Gen. Emmons -- Hawaii   RAdm. Spruance -- Chief of Staff
Admiral Nimitz -- CinC Pacific Fleet   Lt. General Buckner -- Alaska

Chester W. Nimitz -- Commander in Chief of Pacific Fleet. Assigned to run the Pacific War soon after Pearl Harbor. Successfully brought victory !

Delos C. Emmons -- assigned to run the Hawaiian Department, Army and Air Force, soon after Pearl Harbor. Later headed Western Defense Command.

Frank J. Fletcher -- Commanded the first three of five great carrier battles of WWII sinking six enemy carriers in the first nine months of the war. Later commanded the North Pacific and accepted surrender of the Japanese northern fleet.

Raymond A. Spruance -- Commander of cruisers of VAdm Halsey's task force until Midway when he became a task force commander under Fletcher, then became Nimitz's chief of staff. Later, alternated with Halsey as commander of the Fifth Fleet.

Simon B. Buckner -- At the start of the war was in charge of Alaska. Later commanded 10th Army and the invasion of Okinawa. Highest ranking officer killed in WWII.

Reference

Ranks, Ratings, and Awards of WWII

Ranks: Officer

O- Officer – Typical duties - Sleeve, Shoulder. Years required for rank

O-1 . Ensign -- assistant - single stripe, gold bar. Newly commissioned.
O-2 . Lieutenant, Junior Grade -- assistant - 1-1/2 stripes, silver bar. 1-3 years of service
O-3 . Lieutenant -- Skipper of patrol boat -- two stripes, two silver bars. 3.5-7 years
O-4 . Lieutenant Commander -- Skipper of destroyer or submarine - 2-1/2 stripes, gold leaf. 6-14 yrs
O-5 . Commander -- Executive officer on capital ship - three stripes, silver leaf. 12-21 yrs
O-6 . Captain -- Skipper of Cruiser and larger - four stripes, eagle. 18-25 yrs
O-7 . Commodore (*wartime only*) -- Convoy commander; same as Rear Adm, lower division, broad stripe, one star
O-8 . Rear Admiral, -- Task Force Commander - broad plus 1 stripe, two stars
O-9 . Vice Admiral - Fleet Commander - broad plus 2 stripes, three stars
O-10. Admiral -- Theater Command, Atlantic/Pacific/Far East - broad plus 3 stripes, four stars
O-11. Fleet Admiral (*wartime only*) -- Worldwide, later awarded to two others. - broad plus 4 stripes, five stars. Only Leahy, King, then Nimitz, and just after the war, Halsey.

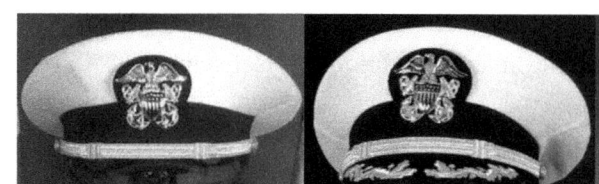

Junior Officer   Commander/Captain   Admiral

Line Officers

General command path. Wears a star on sleeve.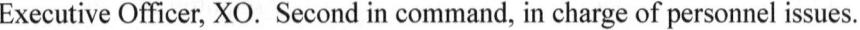
Captain, CO. Officer in command of the ship.
Executive Officer, XO. Second in command, in charge of personnel issues.
Drill Officer, DO. Third in command, in charge of training and discipline.
First Lieutenant. In charge of cleanliness and upkeep.
Maintenance Officer. In charge of cleanliness and upkeep ashore.

Operations Officers.

Navigator. -- responsible for safe movement from one place to another.
Gunnery Officer. Responsible for ordnance equipment and training.
Engineering Officer. Responsible for machinery for driving the ship
Damage Control Officer Responsible for water-tight integrity, ships, boats and structural parts of the ship.
Watch Officer, OOD -- Officer Of the Deck, responsible for conditions during a watch period.

Staff Officers

Staff Corps officers wear their specialty insignia on the sleeve of the dress blue uniforms and on their shoulder boards in place of the star worn by Line officers. On Winter Blue and khaki uniforms, the specialty insignia is a collar device worn on the left collar while the rank device is worn on the right.

Chaplain -- religious and welfare interests.

Civil Engineering - construction

Judge Advocate - legal

Medical / Dental / Nurse -- health and conditions- 7 leaf symbol

Supply -- purchase and disbursement-- 3 leaf symbol

Reference

Officer Numbers and Average Age and Service Time

Rank	Average Age	% of Officers	% Selected	Years in Grade
O1 Ensign	21-23	16%-18%		
O2 Lieutenant, Jr Grade	24-26	16%-18%	92%	1
O3 Lieutenant	28-30	21%-23%	92%	3
O4 Lieutenant Commander	35-37	17%-19%	90%	6
O5 Commander	41-43	12%-14%	81%	6
O6 Captain	46-48	5%- 8%	45%	6
O7-O10 Admirals		0.5%-0.7%		

Officers comprise about 10% of total Navy on active duty. Warrant officers add another 1-2%

Separate officer lists are maintained for special branches : medical, chaplain, flag officers, warrant officers, WAVES, etc.

Officers selected are placed on a promotion list and promoted as vacancies occur in theorder of date of rank.

Temporary promotion in rank is granted when large numbers of officers are needed, as in war time, which will accelerate the promotion timetable.

Exceptional officers may be promoted early, 5 to 10% of promotions are on such "fast track".

An officer review board sets dates of rank to be considered for each grade and the method of selection, based on the needs of the service. The "fully qualified" method promotes all who are not unqualified. Fully qualified is generally used for Lt(jg) and Lieutenant grades. The "best qualified" method selects a percentage of those most qualified. Best qualified is used for grades of Lieutenant Commander,, Commander, and Captain.

Warrant Officers

Technical Specialists (without college)

W1 -Warrant Officer, non-commissioned, half stripe with blue
W2- Chief Warrant Officer, commissioned, one stripe with blue. 1-2% of officer corps.

Midshipmen -- Officers in Training

Ratings: Enlisted Ranks

E - Enlisted. See USN for today's ranks, only E1-E7 existed in WWII.

E1 . Seaman 3rd class (now Seaman Recruit) - while in Boot Camp; 4 month
E2 . Seaman 2nd class (now Seaman Apprentice) - completed boot camp, on first posting.
E3 . Seaman 1st class (now Seaman) - passed an examination on Navy and seamanship
E4 . Petty Officer 3rd Class - qualified to assist with a specialty
E5 . Petty Officer 2nd Class - capable in most areas of specialty
E6 . Petty Officer 1st Class -- knowledgeable in all areas of specialty.
E7 . Chief Petty Officer -- knowledgeable in all areas of specialty and rating with a special area of competence (leadership).

Rates, Jobs

Seaman -- deck, ordnance, electronic, clerical
Fireman -- engineering, black gang, left arm rate
Airman -- aviation
Constructionman -- SeaBee
Hospitalman/Dentalman -- medical
Steward -- officer servant
Yeoman – clerk

Left arm rate -- engineering machinery and hull specialist, typically below decks.
Right arm rate -- deck and operations specialist.
Boatswain -- deck crew specialist
Coxswain -- 3rd class boatswains mate, small boat helmsman.

All petty officers may be called by a customary nickname.
Boatswain's Mate -- "Boats"
Radioman -- "Sparks" or "Sparky"
Gunnery Mate -- "Gunner" or "Gunny"
Machinists Mate -- "Chips"
Any CPO – "Chief"

Ships Bells. A bell is rung every half hour, adding one each time, until the end of a four-hour watch is marked with eight bells.

Awards of WWII

Award		Points
Good Conduct Medal	3 years with minimum offenses.	1
Purple Heart	Received medical attention in combat zone.	1
Commendation Ribbon	Letter of commendation from Sec of Navy.	2
Air Medal	Meritorious achievement in flight.	3
Bronze Star	Meritorious achievement in a combat zone.	3
Navy & Marine Corps Medal	Heroism not involving combat.	3
Distinguished Flying Cross	Heroism in flight.	4
Legion of Merit	Exceptionally meritorious conduct, war or peace.	4
Silver Star	Conspicuous gallantry in combat.	4
Distinguished Service Medal	Exceptionally meritorious in a position of great responsibility, combat or peace.	4
Navy Cross	Extraordinary heroism in combat.	4
Medal of Honor	Conspicuous gallantry, risk of life above and beyond.	5

Devices on ribbon

"v"	Valor, the award was earned in combat.
"stars"	Gold star for subsequent awards. 5 gold stars = 1 silver

Home Front in 1941

Life in the United States in the year entering World War II

The Great Depression began to abate by 1939 with the rumors of war in Europe. There was a double influx of business, first from European countries that had a need to supplement their arms and commodities, and second from an increase in military spending in the United States. By mid-1941 the U.S. was in the Atlantic war for all but a formal declaration.

GOVERNMENT
President : Franklin D Roosevelt ;   Vice Pres : Henry A. Wallace
Speaker of the House : Sam Raburn ; President of the Senate: Wallace
Cabinet: Sec State : Cordell Hall ;   Treasury : Henry Morgenthau, Jr.
War : Henry L. Stimson ;   Attorney General : Francis Biddle
Postmaster General : Frank Walker ;   Navy : Frank Knox
Interior : Harold L. Ickes ;   Agriculture : Claude R. Wickard
Commerce : Jesse Jones ;   Labor : Frances Perkins
Chief Justice : Harlan F. Stone
U.S. Supreme court upholds minimum wage law. 25¢/hr.
National Gallery of Art opens.
U.S. Savings Bonds introduced.
Office of Price Administration, OPA, formed.
Justice Louis Brandeis (1856-1941) dies.

Census	1940	2000
Population:	131,669,275	281,421,906.
Farms:	6,102,000 .	2,172,000.
Farm population:	30,547,000 ; 23.2%	4,801,000 ; 1.9% (1990)
Home Ownership:	43.6%	67.4%
Immigration:	528,431	7,605,066
% of growth	7.4%	37.0%
EPA Emissions:		
Carbon monoxide:	94MM tons	97MM
Sulfur dioxide :	20MM tons .	19MM
Public Education:	$2,261 MM	$325,976 MM
Life Expectancy:	Male 65 ; Female 70	Male 74 ; Female 80
Infant Mortality:	47 /K	7 /K
Alcohol Use:	1.56 gallon per capita	2.2
Cigarettes:	1,976 per capita	2,103
Newspapers:	1,878	1,480
Road Mileage:	3,287,000 paved, 13%.	3,932,000 paved, 62%.

Population Mix – U.S. Census

Mother Tongue	1940	2010
English	78.59%	79.41%
German	4.16	0.37
Italian	3.18	0.25
Polish	2.17	0.21
Spanish	1.57	12.79
Yiddish	1.48	0.05
Scandinavian	1.45	0.05
French	1.19	0.46
Asian		2.68
India		0.90
Other	2.83	2.49
Race		
White	90%	64%
Negro	10%	12%
Other	< 0.5%	24%
(Hispanic)	N/A	(16%)
Church Members	55.8M = 42%	182.2M = 79.8%
Roman Catholic	36%	31%
Baptist	15%	20%
Methodist	13%	6%
Lutheran	12%	5%
Jewish	8%	2%
Presbyterian	8%	3%
Latter-day Saints	4%	2%
Episcopal	2%	1%
Other Christian	2%	27%
Islam	0.8%	1%
Other	<0.5%	3%

1943 World Almanac and Book of Facts.
2012 World Almanac.

Jewish   US = 2.3% ; NY = 17% ; NYC = 28% ; Bronx = 44%.
Japanese   1943 = 0.1% ; citizen = 63% ; alien = 37%.
               CA = 71.0% ; WA = 11.5% ; OR = 3.1% ; NY = 1.9% ; other = 12.5%

Most popular radio shows : The Red Skelton Show ; Inter Sanctum ; The Great Gildersleave ; Duffy's Tavern ; The Thin Man.

Movies :
   Best Picture   : *How Green Was My Valley*
   Best Director : John Ford , *The Grapes of Wrath*
   Best Actor   : Gary Cooper, *Sergeant York*
   Best Actress   : Joan Fontaine, *Suspicion*
   Supporting Actor : Donald Crisp , *How Green Was My Valley*
   Supporting Actress : Mary Astor, *The Great Lie*

Popular Pictures : *Citizen Kane* -- Orsen Welles
   The Maltese Falcon - Humphrey Bogart
   Gone with the Wind ; *Cheers for Miss Bishop*;
   Fantasia ; *The Wizard of Oz*

Top Movie Names : Mickey Rooney ; Clark Gable ; Abbott and Costello ; Bob Hope ; Betty Davis ; James Cagney ; Judy Garland.

Songs : Bewitched, Bothered, and Bewildered ; Deep in the Heart of Texas ; There, I've Said It Again ; Racing with the Moon ; By the Light of the Silvery Moon ; Chattanooga Choo Choo ; I Don't Want to Set the World on Fire ; I Got It Bad and That Ain't Good.

Bands : Stan Kenton ; Benny Goodman ; Glenn Miller.

Singers : Frank Sinatra ; Bing Crosby

Broadway : Arsenic and Old Lace

Pulitzer Prize in Journalism : St Louis Post-Dispatch for city smoke nuisance.

Pulitzer Prize for Drama : There Shall Be No Night - Robert Sherwood.

Best Sellers :Fiction - For Whom the Bell Tolls ; How Green Was My Valley.
   Non-fiction - I Married Adventure ; Berlin Diary.

Poets : Edna St. Vincent Millay ; e e cummings

Economics	**1940**	**2010**
Federal spending :	$14 billion	$ 3,555 billion
Federal debt :	$57 billion	$13,050 billion
Consumer Price Index :	14.7	218.7
Unemployment :	9.9%	9.8%
Cost of a first-class stamp :	3 cents	44 cents

Reference

SPORTS
World Series : NY Yankees vs. Brooklyn Dodgers (4-1)
Stanley Cup : Boston Bruins vs. Detroit Red Wings (4-0)
Wimbledon : Men : *Not held*. Women : *Not held*
Kentucky Derby Champion : Whirlaway (Triple Crown!)
NCAA Basketball Championship : Wisconsin vs. Washington St. (39-34)
NCAA Football Champions : Minnesota (8-0-0)
Grand Slam : Men - Bobby Riggs ; Women - Sarah Palfrey Cook
The Masters : Craig Wood
Heavyweight : Joe Louis
Indianapolis : Floyd Davis , Mauri Rose
Joe DiMaggio safely hit in 56 consecutive baseball games.
Lou Gehrig (1903-1941) died, baseball player.

ENTERTAINMENT
"Citizen Kane," Orson Wells. Using newly developed film stocks and a wider, faster lens, he pushes the boundaries of montage and setting as well as sound, redefining the medium.
Actress Greta Garbo retires at age 36.
Paderewski (1860-1941) Polish composer died.
Miss America : Rosemary LaPlanche (CA)
Walt Disney's animated film *Dumbo* is released.

SCIENCE
Nobel Prizes:
    Peace: *None awarded.*
    Chemistry: *None awarded.*
    Physics: *None awarded.*
    Physiology & Medicine: *None awarded.*
Glenn Seaborg and Edwin McMillan isolate plutonium, a fuel that will be preferable to uranium for nuclear reactors.
RCA demonstrates a new electron microscope that magnifies 100,000 times.
John Rex Whinfield and Dickson invent Dacron.
Construction begins on Gatun Locks, Panama Canal.
Grand Coulee Dam, Washington, begins operation.
Rainbow Bridge, Niagara Falls opens.
"Bug bomb" aerosol container developed.
John Vincent Atanasoff creates the first modern digital computer.
Z3 computer containing 2,600 relays is first programmable computer.
Caproni-Campini fan-jet plane flies Milan to Rome at 310 mph.
Lord Baden-Powell (1857-1941) died, founder of Boy Scouts.
NBC and CBS granted TV licenses in NYC to use commercials.

CHRONOLOGY 1941

March 1941
11. The Lend-Lease Act of 1941 passed by the U.S. Congress, gave the President power to sell, transfer, lend, or lease such war materials. Total lend-lease aid exceeded $50 billion, of which the British Commonwealth received some $31 billion and the USSR received over $11 billion.

May.
15. Meteor, first British jet aircraft, used Whittle engine.

June
20. Book, "Berlin Diary" released.
30. Arms production : 2 battleships, 24 destroyers, 9 submarines; 151 fighters, 70 bombers; 260 light tanks, 130 medium, zero heavy tanks.

July
1. Chrysler, Ford, Hudson, Nash, and Studebaker have all announced price increases ranging from $10 to $53 per car. The government tried to talk them out of it.
5. RAF Cadets start 30-week course of flight training at U.S. airfields.
7. FDR announces secret landing of U.S. troops in Iceland to prevent Germany from taking strategic outposts in the Atlantic. *(Actually, Marines release British troops for combat elsewhere.)*
28. Fifty oil tankers had already been turned over to Britain. The oil industry was asked by the Petroleum Coordinator, Harold Ickes, to turn over 100 more tankers. *(This will leave only 200 oil vessels under the U.S. flag.)*

August
1. 19 million man-days lost to strikes in the last year.
11. The U.S. government becomes sole buyer/seller of raw silk. Silk is needed for parachutes and ammunition powder bags. Causes rush to stocking counters. A new product, nylon, tougher than silk and also sheer was introduced and quickly accepted.
18. Steel was put on mandatory priorities; no steel can be sold without the purchaser having a government priority. Capacity is only 90% because of a shortage of scrap metal. *Over 8 million tons had been sent to Japan.*
22. U.S. Army doubles in last year since the President called 60,000 National Guardsmen for 12 months service ; a recruit earns $21 per month, to $30 after 4 months. Chief Petty Officer $126.

September
1. Battle damaged British warships repaired in U.S. naval yards.
5. USS *Greer* attacked while carrying mail to Iceland. *(Actually, Greer was tracking U-652 and each attacked the other with no effect.)*
11. President warns Axis ships against entering American waters. *(Actually, a "shoot on sight" order to USN.)*
11. Ground was broken for the Pentagon office building.

Reference

13. Selective Training Act (draft) signed for men age 21-35 for one year of training service. *(Actually, the target for invasion of Europe was May 1943.)*
30. Three U.S. merchant ships sunk this month: *Steel Seafarer, Sessa;* and *Montana.*

October
Willys-Overland "Jeep" (general purpose vehicle) selected by Army.
16. Draft registration of 17,000,000 men.
17. U.S. destroyer delivering mail is torpedoed near Iceland. *(Actually, USS Kearny was escorting a convoy attacked by a U-boat "wolf pack".)*
16-18. Japanese cabinet resigns ; General Tojo becomes Prime Minister.
29. Draft drawing to select 800,000 men for service.
31. U.S. destroyer **Ruben James** sunk by a torpedo while a convoy escort.

November
7. U.S. Senate revokes Neutrality Act. House follows on Nov 13.
13. Draft age lowered from 21 to 18.
16. CIO coal strike.
18. Japanese meet to negotiate in Washington, DC.
25. U.S. troops to Dutch Guiana to protect bauxite mines.

December 1941.
1. State of emergency proclaimed in Singapore and Malay States.
8. FDR speech, "A Date that Will Live in Infamy."
8. Chicago Tribune says U.S. battle fleet is on its way to Japan.
(Actually, the battle fleet was sitting on the bottom of Pearl Harbor.)
10. Two Japanese battleships reported sunk in invasion of Philippines. *(Actually, no Japanese warships larger than a destroyer were lost in the first 5 months of the war.)*
11. Germany-Italy and USA declare war against each other.
13. Japanese invasion of Philippines thrown back. *(Actually, 80,000 Japanese troops were landed over a two week period against 10,000 American troops.)*
15. Naval Secretary Knox reports 2 battleships and 3 destroyers lost at Pearl Harbor. *(Actually, 5 battleships were sunk and 3 needed months of repair.)*
17. MacArthur reports Philippine situation well in hand. *(Actually ...)*
20. Japan's new Zero fighter said to closely resemble Brewster Buffalo. *(Actually, the Zero is superior to all existing world fighters including the Buffalo and its successor, the Grumman Wildcat.)*
22. It is revealed that PM Churchill and Chiefs of Staff are in Washington.
23. Union no-strike pledge.
25. North Platte Canteen opens - serves sandwiches and coffee to 6,000,000 in uniform – until April 1, 1946

By year-end San Francisco had its tenth alert, 7th blackout.

Enemy Alien Relocation
Timeline

1880. Census shows only 148 Japanese in the whole of USA.
1882. Chinese Exclusion Act ; is silent on Japanese.
1890. Census shows 2,000 Japanese.
1910. Census shows 72,000 Japanese have replaced Chinese workers. Mostly young men.
1913. California Alien Land Law
1924. Japanese Exclusion Act.
Sep 19, 1931. China invaded ; puppet state of Manchukuo (Manchuria) formed by Japanese Army.
Mar 27, 1933. Japan withdraws from League of Nations.
June 7, 1937. Start of Sino-China war.
Dec 13, 1937. Rape of Nanking: 200,000 Chinese murdered. Shocks world.
Sept 1, 1939. Germany invades Poland. War declared in Europe on Sept 3.

----- 1940 -----

Census showed 126,947 Japanese ; 40,000 now over 50 years as head of household, mostly on West Coast.
- Issei - born in Japan, enemy aliens.
- Nisei - born in U.S. or Hawaii, dual citizens by birth.
- Kibei - born in U.S., schooled in Japan ; also, any preferring Japan to the U.S.
- 30% of Nisei held Japanese citizenship and 60% of Issei had not renounced Japanese citizenship.

Sep 27. Japan, German and Italy sign Tripartite Pact to support each other's military aims.

----- 1941 -----

Spy, LCdr Ohmea, allowed to return to Japan so as to keep tabs on his successor.
May 19. Spy Nakauchi message intercepted : *"We have already established contact with absolutely reliable Japanese in San Pedro and San Diego area who will keep a close watch on all shipments of airplanes and other war material . . ."*
June 7. Spies LCdr Itaru Tachibana and Toraichi Kono arrested.
Dec 6 . Spy Yoshikawa reports ship stations at Pearl Harbor to Tokyo ; 36-minutes later is decyphered aboard *Akagi*.

December 7, 1941. Pearl Harbor attacked.
Dec 7 . Malaya and Siam invaded.
Dec 7 . The Philippines attacked.
Dec 7 . Wake Island attacked, repulsed. Two Japanese destroyers shell Midway.
Dec 7 . FBI arrested 1,300 aliens, including those implicated by the Tachibana/Kono ring.

Reference 235

Dec 8. Japan takes the Gilbert Islands
Dec 8. HMS Prince of Wales *and* Repulse *sunk off Malaya.*
Dec 11. Burma invaded.
Dec 12. Downed Zero pilot with aid from local Japanese-American (Nisei) takes Hawaiian village hostage.
Dec 16. Borneo invaded.
Dec 20. Japanese submarine I-17 sinks U.S. tanker Emidio *off Cape Mendocino, California.*
Dec 22. The Philippines invaded.
Dec 23. Japanese submarine I-21 sinks U.S. tanker Montebello *four miles south of Piedras Blancas light, California. I-21 machine-guns the lifeboats, but inflicts no casualties.*
Dec 23. Japan takes Wake Island
Dec 25. Enemy diplomats, including 540 Japanese, moved into resorts: Greenbrier at White Sulphur Springs, WV ; The Homestead at Hot Springs, VA ; and Park Grove Inn in Asheville, NC.
Dec 25. Hong Kong surrenders.
Dec 31. Japan occupies Manila.

----- *1942* -----

Jan 11. *Dutch East Indies invaded. The entire white male population of Balikpapan is slaughtered.*
Jan 23. Intercept : "The brilliant success of our armed forces at Pearl Harbor was due, mainly, to the military information based on reports sent by our informers on the spot, whose efforts represent untold sacrifices in blood and tears."
Jan 29. Enemy aliens excluded from strategic military and industrial areas; includes Japanese populated coastal fishing and farming villages.
Feb 2 . Sheriffs & D.A.s meeting calls for evacuation for U.S. security and for the safety of Japanese.
Feb-March. Vigilantes turn back Japanese trying to leave California.
Feb 9 . SS Normandie *burns at New York and capsizes.*
Feb 10. Japanese given 48-hour deadline to settle affairs and leave coastal zone.
Feb 15. Singapore surrenders; mass rape of white women.
Feb 19. FDR executive order 9066 excluding Japanese from the entire coastal area.
Feb 19. Japanese air raid on Port Darwin, Australia. Port is abandoned.
Feb 20. Japanese invade Timor Island, N.E.I., 350 miles from Australia.
Feb 21. Congressional hearings on relocation, Toland Committee.
Feb 23. Japanese submarine I-17 shells oil tank farm in California.
Feb 24. *Air Raid alarm at Los Angeles; 1,400 3" AA shells fired at an errant weather balloon.*
Feb 24. Canada enacts a 100 mile coastal exclusion zone.
Feb 27. Battle of Java Sea - defeat of Allied navies in the Far East.

March 1-5. Eleven retreating USN ships of Asiatic fleet sunk by IJN.
March 2 . West Coastline declared restricted zone, Japanese told to move to the interior.
March 9 . Java surrenders.
March 13. Solomon Islands invaded by Japanese.
March 18. War Relocation Authority formed.
    Coupon books issued to the relocated.
March 22. First contingent from Los Angeles to Manzanar Assembly Center.
March 29. End of voluntary relocation period.
April 5 . Imperial Japanese Navy raids British Eastern Fleet at Ceylon; sink carrier, two cruisers, a destroyer, and 24 ships of the merchant fleet.
    5,000 Germans relocated to Ellis Island.
April 9 . U.S. Army surrenders Bataan, Philippines. Bataan death march.
April 18. Doolittle Raid on Tokyo.
April 19. Japanese land in Dutch New Guinea.
April 30. Exclusion order -- to be 70 miles from coast by May 7.
May 1 . Start move to assembly centers. (1)
May 1 . Mandalay, Burma, falls to Japanese Army.
May 7 . European diplomats, 1,600, returned on Swedish ships.
May 7-8. Battle of Coral Sea. USS Lexington *lost but Japanese invasion of south coast of New Guinea turned back.*
May 20. Burma surrenders.
May 21. U-boat attacks on tankers in Atlantic Coast & Caribbean is at height.
June 1 . Manzanar Assembly Center becomes Relocation Center. (2)
June 3 . Japanese carrier force bombs Dutch Harbor, Alaska.
June 4-6. Battle of Midway defeats Japanese capability to invade Hawaii.
June 5-7. Japanese bomb Dutch Harbor, occupy Attu and Kiska, Aleutians, without opposition.
Jun 15. Seven U.S. ships torpedoed or mined by U-boats in Caribbean.
Mid-June. Relocation centers ready for occupation for up to 130,000.
June 18, First repatriation of 1,083 on MS *Gripsholm* through Mozambique.
Jun 20. Japanese submarine I-26 shells Estevan Point, Vancouver Island.
July 1942-October 1943. Period of Balance of Sea Power in the Pacific.
July . U.S. troops to South Pacific to build supply and air bases.
Aug 7 . Evacuation from war zone complete : 110,000 people
Aug 7 . Marines land at Guadalcanal. Fight until Feb 9, 1943.
Aug 9 . Battle of Savo Island, Allied cruiser force wiped out.
August. Nisei who were not Kibei are allowed to resettle in Midlands or East.
Sep 2 . Second exchange on *Gripsholm*, 1,500, through Portuguese India.
Sep 9 . Seaplane from Japanese submarine *I-25* drops incendiary bombs near Mount Emily, Oregon, to ignite forest fires.
Sep 24. Japanese expand in Gilbert Islands.
Sep 26. Japanese invasion of Port Morseby, thrown back Oct 5.

Reference

Sep 29. Seaplane from Japanese submarine *I-25* again bombs forest in Oregon with incendiaries.
Oct 12. Italians excluded from enemy alien restrictions.
Oct 30. Moving from assembly to relocation centers complete.
Nov 8 . Allied invasion of North Africa.
mid-Dec. Germans on West Coast excluded from enemy alien restrictions.

----- **1943** -----
Jan 28. Combat team of Nisei volunteers authorized; activated Feb 1.
Feb 9 . Guadalcanal battle ends.
Feb 10. Loyalty questionnaire to determine military duty for Nisei males over age 17 and for resettlement permission of all others.
Feb 22. Battle of Kasserine Pass. U.S. Army loses in its first battle.
Mar-Oct. Resettlement allowed. 15,000 loyal Japanese resettle in Midwest.
mid-April. 2,700 Nisei recruits arrive in Camp Shelby, Miss, for training.
April 21. Nisei in uniform suspended from exclusion orders.
July 10. Allies invade Sicily.
July 15. Tule Lake, Cal. becomes Segregation Center for disloyal.
mid-Sept-Oct. Transfers to Tule Lake Segregation Center ; 6,000 loyal out.
Sept 8 . Italy surrenders as Allied invasion force approaches.
Nov . U.S. strikes back in Pacific: Bougainville, Makin, Tarawa.
Nov 5 . Tule Lake becomes unruly; Army takes over with martial law.
Nov-Dec. 20,000 more resettle, given a bus ticket and $25, about a week's wage.

----- **1944** -----
January 14. Army returns Tule Lake to WRA administration.
Jan 24-end. 442nd Nisei Regiment in Italy; earns most medals for gallantry.
January 31. Marshall Islands campaign.
March 15. Japanese invade India.
June 6 . D-Day, Allies land at Normandy, France.
June 15-July 9. Marianas campaign: Saipan.
June 19. Japanese carrier fleet defeated in the Philippine Sea.
June 30. Jerome Relocation Center closed. Half had resettled and half moved to other centers.
July 21-Aug 10. Guam, Tinian campaign.
Oct 20. U.S. landing at Leyte, Philippines.
Dec 17. Exclusion order for West Coast rescinded, relocated can go home in January.
Dec 17. Relocation Centers to close in 12 months.

----- **1945** -----

January 2 . Order of exclusion was rescinded. All free to return to West Coast homes ; except those who renounced citizenship : 5,000 males held in Tule Lake and another 5,000 excluded from the West Coast. There were still more than seven months of warfare remaining with Japan.
January 16. End of Battle of the Bulge.
Feb. 19-March 16. Iwo Jima.
March 9 . B-29 firebombs destroy Tokyo.
April 1 . Invasion of Okinawa.
May 7 . **Germany surrenders.**
June 21. Kamikazes have wrecked 300 U.S. ships by the end of Okinawa.
August 6 . Hiroshima destroyed by one bomb.
August 14. **Japan surrenders.**
Oct-Nov . Relocation Centers closed.

----- **1946** -----

March 20. Tule Lake Segregation Center closed.
    4,000 Nisei and Kibei who renounced America are declared aliens ineligible for citizenship.

----- **Notes** -----

(1) Assembly centers are temporary locations of May 1942 to get Japanese away from war zone. These were CCC camps, fairgrounds, race tracks, etc.
(2) Relocation Centers are low-cost housing built in remote, inland areas.

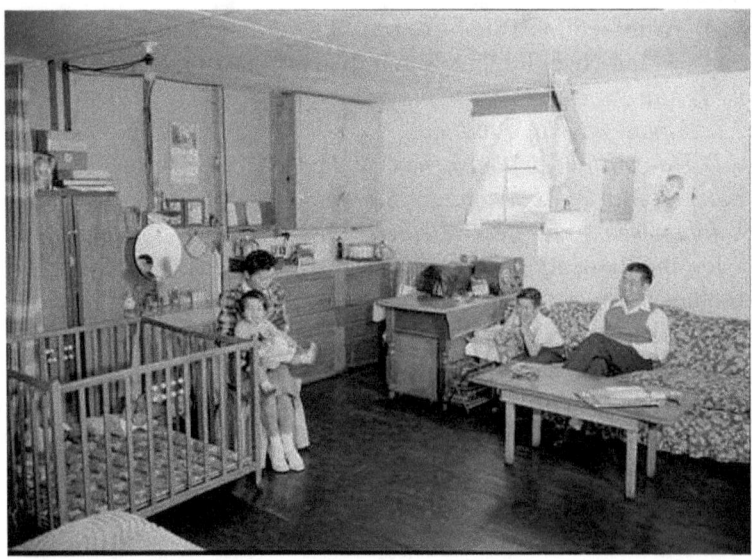

Reference

Relocation Center	Opened	Peak Pop.	Closed
Gila River, AZ	20Jun42	13,348	10Nov45
Granada, CO	27Aug42	7,318	15Oct45
Heart Mountain, WY	12Aug42	10,767	10Nov45
Jerome, AR	01Oct42	8,497	30Jun44
Manzanar, CA	21Mar42	10,046	21Nov45
Minidoka, ID	10Aug42	9,397	28Oct45
Poston (Colorado River) AZ	08May42	17,814	28Nov45
Rohwer, AR	19Sep42	8,475	30Nov45
Topaz, UT	11Sep42	8,130	31Oct45
Tule Lake, CA	27May42	18,789*	20Mar46

* Tule Lake reaches its peak population 25Dec44 as disloyals are concentrated there.

Manzanar, in California's Owens Valley, a former WPA Camp, was designated as an assembly center; the first 80 people arrived Mar 21, 1942, and it was filled by mid-April. When the War Relocation Authority (WRA) was formed, it became the first Relocation Center. Issei, Nisei, and Kibei factions formed ; riots, murders ; then settled down.

Passes, permanent or temporary, were required to leave the relocation centers. Workers and college students, for example, were allowed to work and go to schools anywhere outside the exclusion zone. Earnings and payments were their own. Nursing students received scholarships. About 2,500 evacuees went to Seabrook Farms' New Jersey plant.

Life goes on : birth, marriages, funerals, and graduations occur in the Centers.

Although free to relocate after March 1943, many without family or friends in the Midwest or East preferred the security of the Centers rather than the uncertainty of finding a place to live on their own where they may or may not be accepted.

Tule Lake became Segregation Center after July 15, 1943, to house those who did **not** claim loyalty to the USA.

Diplomats, businessmen and others, 4,700 people, repatriated to Japan in 1942 on neutral ships, most notably Swedish MS *Gripsholm* chartered by U.S. State Department carrying Japanese and German nationals to exchange points where she then picked up U.S. and Canadian citizens. She made 12 round trips, carrying a total of 27,712 repatriates in both Atlantic and Pacific.

442nd Regimental Combat Team – the all-Nisei unit, was the most decorated unit of the war. They are assigned the most difficult tasks in Europe. Consisted of :
   100th Infantry Battalion ; 522 Field Artillery ; 232nd Engineers.

Relocation Center Housing

No, this is American Sharecropper housing

Major source : *America's Concentration Camps* by Allan R. Bosworth.
Other sources : *The Relocation of the Japanese-Americans During World War II* by Dwight Murphey
National Park Service

Loyalty Indicators of Relocated Alien Families

Starting February 6, 1943, internees were given a loyalty questionnaire to determine how to encourage their integration back into society. Two questions were aimed at draft age men. Twenty-two percent of the 21,000 Japanese eligible to register answered "No" to both questions.

> Question 27 "Are you willing to serve in the armed forces of the United States on combat duty and serve wherever assigned?"
>
> Question 28. "Will you swear unqualified allegiance to the United States of America and faithfully defend the United States from any and all attack by foreign or domestic forces, and forswear any form of allegiance or obedience to the Japanese emperor?"

Males between the ages of seventeen and thirty-eight.

"The following figures speak for themselves, showing as they do that about one out of four Nisei in the relocation centers stated their loyalty was to Japan, and that only six out of every one hundred of them considered themselves to be true Americans by volunteering for service in the United States Army."

Relocation Center	Number Registered	Answering "no" to loyalty questions	Number of Volunteers for Army
Central Utah	2,420	806	116
Colorado River	3,356	671	238
Gila River	2,488	547	119
Granada	1,117	117	121
Heart Mountain	1,881	451	47
Jerome	1,341	110	33
Manzanar	1,826	913	101
Minidoka	1,607	32	310
Robwer	1,585	300	37
Tule Lake	2,342	836	59
TOTAL	19,963	4,783	1,181
Average (percent)		**24%**	**6%**

Sources
Double Victory: A Multicultural History of America in World War II
    by Takaki, Ronald
Betrayal from the East by Alan Hynd

These statistics are more understandable once we realize that most of these were Japanese people working in the United States as a place of employment with no commitment to the American Way of Life. Their children born in the U.S. held dual citizenship -- Japanese through their parents and American by way of birth location.

There were Americans living in areas occupied by the Imperial Army, in the Philippines for example, who were captured with no loyalty to Japan. These civilians owned businesses, worked for foreign corporations, were military dependents, or who merely enjoyed the Filipino lifestyle. However, contrast the treatment of American civilians interned by the Japanese.

> "We were put in different internment camps (16-year old girl separated from her store manager parents) and we were able to exist in a very primitive way. We lived on practically nothing. We had to buy our own food because the Japanese kept telling us we were only in protective custody ; they had no reason to feed us, and they did not have to follow the Geneva Convention rules." -- Margaret Gillooly

> "It was terrible to watch the children as the starvation process proceeded. They became like little old men and women. They would sit on the sidewalk and have absolutely no energy to accomplish anything. . . . When Manila was recaptured, they found tons of Red Cross supplies stored in Manila that the Japanese had not released to us." -- Hattie Brantley, a nurse in the same camp.

Contrast this with 4,300 youths from Tule Lake Relocation Center alone that were sent to college on scholarships. Or, that Japanese citizens visiting the United States trapped by the war were awarded $20,000 in 1978 for their inconvenience, including 3,500 returned in a civilian exchange early in the war.

Miscellaneous notes:

The period of forced relocation of less than one year was the same period where civilian rationing brought charges of coddling of Japanese and of POWs who were fed balanced diets while Americans scrimped.

By December 1943, most camps had relocated people working outside the camp as day laborers and farm workers, the same work many had done in the prewar period. Their West Coast jobs were taken by Mexican workers imported to take the place of the Japanese field workers.

In December 1944, the West Coast was declared to be no longer a war zone -- those relocated could return to their homes if they wished. Many did not return until after the war for fear for their lives. The camps remained open for their convenience.

You can estimate the orientation of a speaker by noting if they say that "peaceful Japanese-Americans were imprisoned" or if it is said that "enemy aliens [with their American-born children] were relocated from a war zone."

Secondary education

Seconds for survival

The Rest of the War

1943 ... The End of the Beginning ...
02Jan   Buna taken on N. coast of New Guinea.
31Jan   *Germany surrenders the 6th Army at Stalingrad.*
09Feb   Guadalcanal declared secure.
22Feb   *Kasserine Pass, Tunisia.*
2-4Mar Battle of Bismarck Sea.
26Mar   Battle of the Komandorski Islands, Alaska
18Apr   Yamamoto shot down by P-38s.
11-30May U.S. retakes Attu, Aleutians.
21Jun   New Georgia, Solomons.
5-17Jul *Kursk tank battle, Russia.*
09July *Allies invade Sicily.*
05Aug   *Germany expelled from Russia.*
06Aug   Battle of Vella Gulf, Solomons.
09Sep   *Salerno, Italy, till 17 September.*
01Nov   Marines land on Bougainville, Solomons.
05Nov   *Sara, Princeton* raid Rabaul to protect Empress Augusta Bay.
21Nov   Makin and Tarawa, Gilbert Islands.
26Dec   Cape Gloucester, Solomons.

1944 ... the Long Road to Victory ...
22Jan   *Anzio, Italy, till 23 May.*
29Jan   Army lands in Admiralty Islands.
31Jan   Marines occupy Majuro Lagoon, Marshall Islands.
01Feb. Roi, Namur, and Kwajalein invaded.
07Feb   Kwajalein, Marshalls, secured.
13Feb   *Menachem Begin begins to blow up British in Palestine.*
18Feb   Massive naval air raid on Truk, Carolines.
19Feb   Marines land on Eniwetok Island, Marshalls.
15Mar   Japan invades India from Burma.
22Mar   U.S. lands at Hollandia, Dutch New Guinea.
04Jun   *Allies enter Rome.*
06Jun   **D-Day**, *Normandy invasion.*
15Jun   B-29s bomb Japan from China.
15Jun   Saipan, Marianas, till 9 July.
19Jun   Japanese carrier fleet defeated in Philippine Sea, "Turkey Shoot."
21July-10Aug Guam.
24July-01Aug Tinian.
25July *Breakout from Hedgerows of Normandy.*
15Aug   *Allies invade southern France.*
22Aug   Japanese retreat from India.

25Aug *Liberation of Paris.*
15Sep-13Oct Invasion of Peleliu, Palaus.
17Sep-26Sep *"A Bridge Too Far", Arnham, Holland.*
20Oct U.S. landing on Leyte, Philippines.
23-26Oct Japanese fleets destroyed at Leyte Gulf.
24Nov Start B-29 raids on Japan from Tinian, Marianas.
15Dec Landing on Mindoro, the Philippines.
16Dec *Battle of the Bulge.*

1945   ... the War is Concluded ...
09Jan Landing on Luzon
16Jan *Battle of Bulge ends; Allies link up.*
4-11Feb Yalta. USSR to enter Pacific War.
13Feb *Firebomb Dresden.*
19Feb-16Mar Iwo Jima.
03Mar Manila liberated.
07Mar *Remagen Bridge.*
09Mar Fire bomb Tokyo.
21Mar Mandalay, Burma, regained.
22 Mar *Patton and Montgomery cross Rhine.*
22Mar *Arab League declares war on Axis.*
01Apr-21Jun Okinawa.
12Apr FDR dies, Truman President.
30Apr *Hitler suicide.*
03May Rangoon, Burma, freed.
07May **Germany surrenders.**
10June Australians invade Borneo.
30June Luzon declared secure *[fighting continued]*
16July Atomic bomb test Alamogordo, NM.
06Aug Hiroshima.
08Aug USSR invades Manchuria.
09Aug Nagasaki.
14Aug **Japan surrenders.**
02Sep Signing on USS *Missouri.*
- - - - - - -

20Sep45 Gandhi and Nehru demand British troops leave India.
01Nov45 Planned invasion of Kyushu, Japan.
01Mar46 Planned invasion of Honshu, Tokyo Plain.
04July46 Philippine independence.
24Sep49 USSR atomic bomb.
25June50 North Korea attacks South Korea.
08May54 Dien Bien Phu, Vietnam.

PLANS for the INVASION of JAPAN

Operation Downfall, the invasion of Japan, was in two components scheduled for the Fall and Spring of 1945-46:

Operation Olympic, Nov 1, 1945, after the hurricane season, before winter. The Sixth Army, with nine divisions (3 more in reserve) was to invade three beaches in southern Kyushu, the southern-most of the four Japanese home islands. This was to become a giant airbase to support the next invasion phase in the Spring of 1946. The Japanese had correctly predicted our invasion point and had reinforced Kyushu threefold over initial U.S. expectations with nine American divisions to attack three of Japanese defenders. The final plan had 18 U.S. divisions attacking 11 IJA divisions in defensive positions.

Operation Coronet, March 1, 1946, invasion of Honshu, the main island, with 22 divisions in the Spring after airfields on Kyushu allowed land-based air support in the conquest of Tokyo.

U.S. Preparations

The previous phase of the war had been the capture of the Marshalls – Saipan, Tinian and the US island of Guam during June, July and Aug 1944. These were captured to provide airfields within the effective range of B-29 Superfortress, very heavy bombers. B-29 attacks started in November 1944, by March 1945, Tokyo, Osaka and all other industrial cities had been bombed.

Air dropped mining began in March 1945, in the Shimonoseki Strait, separating Kyushu and Honshu, to isolate the invasion island. (The USAAF had run out of bombs.) Over 120 ships succumb to these mines. Submarine efforts were concentrated in the Sea of Japan on the northwest coast while carrier task forces concentrated on the Pacific Ocean side.

Carrier Task Forces. The first strike on Japan's home islands was the period 18-22March 1945 to disrupt attacks on our invasion fleet as it approached Okinawa. Raids by 11 fleet carriers and 6 light carriers destroyed aircraft such that the Japanese air attacks on U.S. forces at Okinawa were delayed until 6 days after the landings.

Two fleets were to participate in Olympic: The Strike fleet with 21 carriers, 10 fast battleships, and their train. The Assault fleet had 1,500 transports and 800 warships including 26 carriers and 13 battleships.
The capture of Kyushu was to be made by Pacific theater troops. Except, USAAF airbase building services were transferred directly from Europe to the Pacific beginning July 1945. The Royal Navy Pacific Fleet had already become Task Force 57 attached to the U.S. Fifth Fleet.

Japanese Defenses

Troops. Japan was scrapping the bottom of a big barrel. Two million new recruits were called up and experienced Armies was brought back from China and Manchuria to defend the homeland.

Kamikaze. Numbers of about 2000 Navy and 3500 Army airplanes have been cited as available for the defense, and of course, preparations would have continued with 500 mini-subs under construction, specially designed aircraft building, motorboat and manned torpedo stations established. The Japanese military was committed to and was convinced they could repel the initial assault. 1,465 Kamikaze had attacked at Okinawa, 400 miles away, had sunk or damaged 250 warships. – a ratio of 1 hit per 6 attempts. Troopships sailing into waters adjacent to Japan, they thought, didn't stand a chance. U.S. planners estimated 250 hits; Japanese planners expected 480 ships sunk. A troopship carries about 1,000 men.

Expected casualties. By this stage in the war, the overwhelming American material condition had reduced the ratio of American killed vs. enemy. The assault by Pacific trained Army troops from the Philippine Campaign and combat hardened Marines lessened the expected causalities on the American side. Conversely, first-rate Japanese troops with pre-war combat experience in China -- which had made the initial conquests in the Pacific against inexperienced Allied troops -- had mostly been killed. The combat trained troops in China had been replaced with secondary troops -- these now experienced troops were recalled to defend the home islands. These troops from China had never been exposed to the massive air attacks that were now normal operations from U.S. land and sea forces. Japan's naval ships had been destroyed.

Prospects of Operation Olympic. Japan fully expected to be able to repel the first landing with the help of suicide tactics. As shown at Normandy, the Americans expected to overpower all in their way. The U.S. expected to have air superiority, which places an imposition on the defense. The Navy expected to interdict all movements of resupply and reinforcement. Best guess, the attack would be a repeat of "bloody Omaha beach" with a successful American landing. The plan called for sealing the mountains rather than fighting an Okinawa type campaign. The goal of establishing air bases would proceed as an American specialty. There would be continued casualties but the goal accomplished.

SOME KEY SPEECHES

Subject	Who	Date	Where
14-Parts	Emissary	Dec 7, 1941	D.C. during Pearl Harbor Attack
A State of War	Roosevelt	Dec 8, 1941	Before the Congress
I Shall Return	MacArthur	Mar 12, 1942	On arrival in Australia
I Have Returned	MacArthur	Oct 20, 1944	On wading ashore
A-Bomb	Truman	Aug 9, 1945	Announcing the attack
Surrender	Hirohito	Aug 15, 1945	Recording played on radio
The End	Truman	Aug 15, 1945	Announcing the end of war
Victory	Halsey	Aug 15, 1945	Statement to the Third Fleet
Signing	MacArthur	Sept 2, 1945	Aboard USS Missouri
To the Fleet	Nimitz	Sept 2, 1945	Upon the Signing
Northern Forces	Fletcher	Sept 8, 1945	Mitsu Wan Harbor

The 14-Part Message delivered an hour after the strike on Pearl Harbor began,

The Japanese Government regrets to have to notify hereby the American Government that in view of the attitude of the American Government it cannot but consider that it is impossible to reach an agreement through further negotiations.

President Roosevelt addresses the Congress of the United States.

Yesterday, December 7, 1941, a date which will live in infamy, the United State of America was suddenly and deliberately attacked by naval and air forces of the Empire of Japan.

The United States was at peace with that nation and, at the solicitation of Japan, was still in conversation with its government and its emperor looking towards the maintenance of peace in the Pacific. Indeed, one hour after the Japanese air squadrons had commenced bombing in Oahu, the Japanese Ambassador and his colleague delivered to the Secretary of State a formal reply to a recent American message. While the reply stated that it seemed useless to continue the existing diplomatic negotiations, it contained no threat or hint of war or armed attack.

It will be recorded that the distance of Hawaii from Japan makes it obvious that the attack was deliberately planned many days or weeks ago. During the intervening time, the Japanese Government has deliberately sought to deceive the United States by false statements and expressions of hope for continued peace.

The attack yesterday on the Hawaiian Islands has caused severe damage to American naval and military forces. Very many American lives have been lost. In addition American ships have been reported torpedoed on the high seas between San Francisco and Honolulu.

Yesterday the Japanese Government also launched an attack against Malaya. Last night Japanese forces attacked Hong Kong. Last night Japanese forces attacked Guam. Last night Japanese forces attacked the Philippine Islands. Last night the Japanese attacked Wake Island. This morning the Japanese attacked Midway Island.

Japan has, therefore, undertaken a surprise offensive extending throughout the Pacific area. The facts of yesterday speak for themselves. The people of the United States have already formed their opinions and well understand the implications to the very life and safety of our nation. As Commander-in-Chief of the Army and Navy, I have directed that all measures be taken for our defense.

Always will we remember the character of the onslaught against us. No matter how long it may take us to overcome this premeditated invasion, the American people, in their righteous might, will win through to absolute victory.

I believe I interpret the will of the Congress and of the people when I assert that we will not only defend ourselves to the uttermost, but will make very certain that this form of treachery shall never endanger us again. Hostilities exist. There is no blinking at the fact that our people, our territory, and our interests are in grave danger. With confidence in our armed forces -- with the unbounded determination of our people -- we will gain the inevitable triumph, so help us God. I ask that Congress declare that since the unprovoked and dastardly attack by Japan on Sunday, December 7, 1941, a state of war has existed between the United States and the Japanese Empire.

MacArthur's arrival in Australia having escaped from Corregidor.

The President of the United States ordered me to break through the Japanese lines and proceed from Corregidor to Australia for the purpose, as I understand it, of organizing the American offensive against Japan, a primary object of which is the relief of the Philippines. I came through and I shall return.

MacArthur after wading ashore at Leyte.

People of the Philippines. I have returned. By the grace of Almighty God our forces stand again on Philippine soil. The hour of your redemption is here. Your patriots have demonstrated an unswerving and resolute devotion to the principles of freedom. . . . Rally to me. Let the indomitable spirit of Bataan and Corregidor lead on. As the line of battle roll forward to bring you within the zone of operations, rise and strike. . . . For future generations of your sons and daughters, strike ! In the name of your sacred dead, strike ! Let every arm be steeled.

President Truman on the second atomic bomb.

Having found the bomb we have used it. . . .

We have used it against those who attacked us without warning at Pearl Harbor, against those who have starved and beaten and executed American prisoners of war, against those who have abandoned all pretense of obeying international laws of warfare. We have used it in order to shorten the agony of war, in order to save the lives of thousands and thousands of young Americans. We will continue to use it until we completely destroy Japan's power to make war. Only a Japanese surrender will stop it.

President Truman's statement that resistance to the Allies was at an end.

I have received a note from the Japanese government in reply to the message forwarded to that government by the Secretary of State on August 11. I deem this reply a full acceptance of the Potsdam Declaration, which specifies the unconditional surrender of Japan.

Halsey announcement to the Fleet with kamikaze falling on the horizon, the last one by close destroyer fire.

Men of the Third Fleet, the war is ended. You, in conjunction with our brothers in arms, of all services and all branches of all services, have contributed inestimably to this final result. You have brought an implacable, treacherous, and barbaric foe to his knees in abject surrender. This is the first time in the recorded history of the misbegotten Japanese race that they as a nation have been forced to submit to this humiliation. I said in 1942 the Nips were no superman. You have helped write *finis* on that estimate in 1945. Your names are writ in golden letters on the pages of history -- your fame is and shall be immortal. Whenever you have met the foe, on the sea, on the land, in the air, or under the water, you have been supreme. Whether in the early days, when fighting with a very frayed shoestring, or at the finish, when fighting with the mightiest combined fleet the world has ever seen, the result has been the same -- victory has crowned your efforts. The forces of righteousness and decency have triumphed. At this moment our thoughts turn to our happy and fortunate homeland, to our loved ones. Deeply rooted in each and every heart is a desire -- now that the tumult and glory of war has ceased and victory -- absolute and unconditional victory -- has crowned our efforts -- to return to our homes.

Hirohito's address his people; His first ever, recorded, played at noon

"We, the Emperor, have ordered the Imperial Government to notify the four countries, the United States, Great Britain, China, and the Soviet Union, that We accept their Joint Declaration. To ensure the tranquility of the subjects of the Empire and share with all the countries of the world the joys of co-prosperity, such is the rule that was left to Us by the Founder of the Empire of Our Illustrious Ancestors, which We have endeavored to follow. Today, however, the military situation can no longer take a favorable turn, and the general tendencies of the world are not to our advantage either.

"What is worse, the enemy, who has recently made use of an inhuman bomb, is increasingly subjecting innocent people to grievous wounds and massacre. The devastation is taking on incalculable proportions. To continue the war under these conditions would not only lead to the annihilation of Our Nation, but to the destruction of human civilization as well."

A Japanese military and a government official sign the document of surrender in Tokyo Bay ; MacArthur concludes ,

We are gathered here, representatives of the major warring powers, to conclude a solemn agreement whereby peace may be restored. The issues involving divergent ideals and ideologies, have been determined on the battlefields of the world and hence are not for our discussion or debate. Nor is it for us here to meet, representing as we do a majority of the people of the earth, in a spirit of distrust, malice, or hatred, but rather it is for us, both victors and vanquished, to serve, committing all peoples unreservedly to faithful compliance with the understanding they are here formally to assume. It is my earnest hope, indeed the hope of all mankind, that from this solemn occasion a better world shall emerge out of the blood and carnage of the past . . . a world dedicated to the dignity of man . . . Let us pray that peace be restored to the world, and that God will preserve it always. These proceedings are closed.

Nimitz broadcast to the Fleet upon the signing of surrender.

On board all naval vessels at sea and in port, and at our many island bases in the Pacific, there is rejoicing and thanksgiving. The long and bitter struggle . . . is at an end. . . .

Today all freedom-loving peoples of the world rejoice in the victory and feel pride in the accomplishments of our combined forces. We also pay tribute to those who defended our freedom at the cost of their lives.

On Guam is a military cemetery in a green valley not far from my headquarters. The ordered rows of white crosses stand as reminders of the heavy cost we have paid for victory. On these crosses are the names of American soldiers, sailors, and marines – . . . – names that are a cross-section of democracy They fought together as brothers in arms; they died together and now they sleep side by side. To them we have a solemn obligation – the obligation to ensure that their sacrifice will help to make this a better and safer world in which to live.

Now we turn to the great tasks of reconstruction and restoration. I am confident that we will be able to apply the same skill, resourcefulness and keen thinking to these problems as were applied to the problems of winning the victory.

On acceptance of surrender of the Japanese North Pacific forces, Fletcher says,

Recalling the rape of Nanking, the treachery of Pearl Harbor, the Death March of Bataan, and the murder, torture, and starvation of our comrades in arms, ours will not be an occupation in the Japanese manner. We have shown the Japanese and the world the superiority of our arms. We must now demonstrate to the world and the Japanese people the superiority of these standards of justice and decency for which we fought.

Fletcher Accepts Surrender of Japanese Northern Forces

Sources

The following books are the most depended upon to start the website that led to this book. I am occasionally both satisfied and embarrassed to find that others - in the hard-copy age - have done a magnificent job of addressing topics that I have more recently compiled as necessary to fill-in my own, and possibly your, understanding of World War II in the Pacific. If you live close to a military historical museum, see if they have a library and join; I wish one were close to me. A point of humor is that I have a web page describing the proper use of citations for students and do not use this knowledge in the following

- **Most Used for References**
  The Pacific War 1941-1945 by John Costello
  History of United States Naval Operations in World War II by Samuel Morison. 15 vols. Most used - 3, 4, 5.
  The End of the Imperial Japanese Navy by Masanori Ito
  And I Was There by Edwin Layton - Chief Pacific intelligence officer
  Victory at Sea by James Dunningan and Albert Nofi
  Atlas of the Second World War by HarperCollins
  Encyclopedia Britannica
  Complete Encyclopedia of World Aircraft, editor David Donald.

- **Most Used Websites**
  Official Navy Chronology
    http://www.ibiblio.org/hyperwar/USN/USN-Chron.html
  Dictionary of American Naval Fighting Ships
    http://www.hazegray.org/danfs/
  Jane's Fighting Ships of WWII - Jane's
  Hyperwar - USN in WWII
    http://www.ibiblio.org/hyperwar/USN/
  Hyperwar: The Official Chronology of the U.S. Navy in World War II
    http://www.ibiblio.org/hyperwar/USN/USN-Chron/USN-Chron-1942.html
  Order of Battle
    http://www.navweaps.com/index_oob/index_oob.htm
  Imperial Japanese Navy
    http://www.combinedfleet.com

Reference *255*

- Other relevant books key to undertaking this project:
American Heritage New History of World War II by Sulzberger & Ambrose
Atlas of 20th Century Warfare by Richard Natkiel
World War II by Ivor Matanle
Long Day's Journey into War by Stanley Weintraub, my English prof at Penn State.
Atlas of the World by NY Times and Times of London
Historic Encyclopedia of World War II, edited Marcel Baudot, *et al.*
Japan's World War II Balloon Bomb Attacks on North America by Robert Mikesh
The shame of Savo: Anatomy of a naval disaster by Bruce Loxton with Chris Clark
"Disaster in the Pacific. New Light on the Battle of Savo Island" by Denis and Peggy Warner.
1942 "Issue In Doubt" Symposium on the War in the Pacific by the Admiral Nimitz Museum. Wayman C. Mullins, ed
Rendezvous at Midway U.S.S. Yorktown and the Japanese Carrier Fleet by Pat Frank and Joseph D. Harrington
Miracle at Midway by Gordon Prange
Japan's Decision to Surrender by Robert Butow
World War II Quiz & Fact Book by Timothy Benford
Illustrated Weapons & Warfare - 24 volumes
The Bluejackets' Manual - U.S. Naval Institute, *my boot camp copy, plus an early wartime edition.*

Other Books, not on the shelf, but are in unlabeled boxes in the machine shed that represent 60 years of reading about WWII. (I have to build some bookcases before they can be brought into the house.) These include : *The Raft, They Were Expendable, Silversides, Thirty Seconds Over Tokyo, Guadalcanal Diary* which comprise the first adult books I ever read (from my parents book-of-the-month subscription during the War) and set a lifelong path to the Pacific Theater.

Of course, there are library copies read over the years which help form one's understanding of The War which shall have to go with titles unremembered.

Since getting serious about this project, the following books have been acquired to contribute to my understanding of the war in its many facets.

AIR

The First Team Vol I&II. by John Lundstrom -- Pacific Naval Air
USN Fighters of WWII by Tillman and Lawson.
McCampbell's Heroes by Edwin P. Hoyt. -- Navy air
The Ragged, Rugged Warriors by Martin Caidin. -- early air
Complete Encyclopedia of World Aircraft, editor David Donald.
The Cactus Air Force by Thomas Miller, Jr. -- Guadalcanal.
Black Cat Raiders of WWII by Richard Knott -- PBY
Fire in the Sky: The Air War in the South Pacific by Eric M. Bergerud.
Flying Fortress by Edward Jablonski -- B-17
The Log of the Liberator by Steve Birdsall -- B-24
Air Facts and Feats compiled by F. Mason and M. Windrow.
Knights of the Air by Ezra Bowen
Aircraft of World War II in Combat, editor Robert Jackson.-- Brief, accurate
   summary of war w/photos.
Wing to Wing, Air combat in China 1943-45 by Carl Molesworth
Point of No Return: The story of the Twentieth Air Force by Wilbur H Morrison
USAAF Fighters of World War 2 in Action by Michael O'Leary -- P-35 thru P-80
America in the Air War by Edward Jablonski
The RAF at War by Ralph Barker
Jane's Fighting Aircraft of World War II

ATROCITY

Knights of Bushido by Lord Russell
Unjust Enrichment by Linda Holmes. -- American POWs
Rape of Nanking by Iris Chang
Last Man Out by Bob Wilbanks -- Palawan Massacre
Unit 731: Japan's Secret Biological War" by Peter Williams and David Wallace

BATTLES

The Lonely Ships by Edwin Hoyt. -- U.S. Asiatic Fleet
Pacific Alamo by John Wukovits -- Wake Island
"1942, Issue In Doubt *: Symposium on the War in the Pacific"* by the Admiral
   Nimitz Museum, edit by Wayman Mullins
Rendezvous at Midway : *U.S.S. Yorktown and the Japanese Carrier Fleet* by Pat
   Frank & Joseph D. Harrington
Miracle at Midway by Gordon Prange
Climax at Midway by Thaddeus Tuleja
Midway; the Battle that Doomed Japan by Fuchida/Okumiya
Unknown Battle of Midway by Alvin Kernan – torpedo squadron
The Battle of Midway: Victory in the Pacific by Richard Hough - short summary
No Bended Knee, The Battle of Guadalcanal by Gen Merrill Twining
Carrier Clash: August 1942 by Eric Hammel -- Guadelcanal and East Solomon
The Thousand-Mile War by Brian Garfield. - Alaska
Storm over the Gilberts by Edwin P. Hoyt -- 1943, Central Pacific
   by James Hornfischer

Reference 257

War at Sea by Nathan Miller
Last Great Victory by Stanley Weintraub -- July-Aug 1945
Pacific Breakthrough by Lawrence Cortesi. -- Marianas - *errors in details*
Great Battles of World War II, vol II: *Pacific Naval Battles* by Pfannes and Salamone
Clash of Carriers by Barrett Tillman - Marianas Turkey Shoot
Valor at Samar by Lawrence Cortesi. - Leyte Gulf
Closing the Circle, War in the Pacific: 1945 by Edwin P. Hoyt
Neptune's Inferno by James D. Hornfischer – Naval Battle of Guadalcanal

DOWNFALL
The Invasion of Japan by John Ray Skates. -- Operation Downfall
Downfall by Richard B. Frank. - Invasion of Japan
Japan's Decision to Surrender by Robert Butow -- intrigue timetable.
The Fall of Japan by William Craig
Codename Downfall: the secret plan to invade Japan by Allen and Polmar

HISTORY
US Navy by Richard Humble
History of the U.S. Coast Guard by Howard Bloomfield.
The Influence of Sea Power upon History by A.T. Mahan
History of U.S. Navy Vol Two, 1942-1991 by Robert W. Love, Jr. -- *weak details*

HOME FRONT
Once Upon a Town: The Miracle of the North Platt Canteen by Bob Green
Don't You Know There's a War On? by Richard Ligeman. -- home front.
Japan's World War II Balloon Bomb Attacks on North America by Robert Mikesh
Page One: The Front Page of History of World War II in New York Times
Washington Goes to War by David Brinkley
The New Dealers' War by Thomas Fleming
The Home Front: U.S.A. by Ronald Bailey, Time-Life
The Greatest Generation by Tom Brokaw

PACIFIC WAR
History of U S Naval Operations in World War II by Samuel Morison. 15 vols.
The Pacific War 1941-1945 by John Costello. Excellence in one volume.
Battle Report - Kari *et al*, Office of SecNavy - 5 vols
The End of the Imperial Japanese Navy - Masanori Ito -- From Japanese side.
Victory at Sea by James Dunningan and Albert Nofi
Jane's Naval History of WWII by Ireland
Time-Life History of WWII
Atlas of the Second World War by Harper Collins
Encyclopedia Britannica
An Illustrated History of WWII edited Hammerton, Vol 5-7
US Army in World War II, Pacific, editor Kent Greenfield, *et al.* 5 volumes
Pictorial History of WWII - Veterans of Foreign Wars - 3 vols
The Pacific Campaign by Dan van der Vat.

Illustrated Story of World War II by Reader's Digest
The U.S. Marine Corps Story by J.Robert Moskin
Dirty Little Secrets of WWII by Dunnigan & Nofi
Military Errors of WWII by Macksen -- *what led to defeats, both sides*
Battle Stations, Your Navy in Action -- many illustrations by USN
The Second World War by Winston Churchill, 6 volumes
American Heritage New History of World War II by Sulzberger & Ambrose
Atlas of 20th Century Warfare by Richard Natkiel
World War II by Ivor Matanle
But not in Shame by John Toland
Rising Sun by John Toland
Atlas of the World by NY Times and Times of London
Historic Atlas of the U.S. Navy by Craig L. Symonds
Historic Encyclopedia of World War II, edited Marcel Baudot, *et al.*
World War II Quiz & Fact Book by Timothy Benford
War in the Pacific edited by Ross Burns
United State Navy in World War II by S.E. Smith
The Longest Battle: The War at Sea, 1939-45 by Richard Hough
Pacific War Papers edited by Goldstein and Dillion
Espionage and Counterespionage: military intelligence by Arch Whitehouse
The Second World War in the East by H. P. Willmott - analysis of enemy position
Flags of Our Fathers by James Bradley – Iwo Jima
How They Won The War in the Pacific: Nimitz and His Admirals by Edwin Hoyt
The Rising Sun by Arthur Zich, Time-Life Books
The First South Pacific Campaign by John Lundstrom

PEARL HARBOR
Pearl Harbor by H.P. Willmott.
Pearl Harbor: Verdict of History by Gordon Prange, *et al.*
Infamy: Pearl Harbor and Its Aftermath by John Toland
Our Call to Arms: The Attack on Pearl Harbor by Life
Military History Quarterly, Vol.4, No.1
Pearl Harbor, Illustrated History by Allan Seiden

PEOPLE
In Bitter Tempest by Stephen Regan -- Admiral Fletcher
Nimitz by E. B. Potter.
Great Admirals of World War II by Pfannes & Salamone.
Wartime Journals of Charles A. Lindbergh -- first hand
Lindbergh and the Battle Against American Intervention in WWII by Wayne Cole
Black Shoe Carrier Admiral: Frank Jack Fletcher, by John Lundstrom
Fleet Admiral King: A Naval Record by King and Whitehill -- meetings

Reference

Yamamoto by Edwin Hoyt
Who Was Who in World War II edited by John Keegan *(Pacific is not his strength)*
A Different Kind of Victory: A Biography of Admiral Thomas C. Hart by James Leutze -- best pre-war
Adm Raymond A. Spruance by VAdm Forrestel
The Days of Fletcher by Andres del Castillo -- Adm Fletcher
Now It Can Be Told: the story of the Manhatten Project by Leslie. R. Groves -- the A-bomb
Military History of World War II by Trevor Dupuy. , 6 vol.
Sea of Thunder by Evan Thomas – Halsey
Legacy of Wings by Frank Smith -- Pitcairn, rotary wings
Maggie of the Suicide Fleet by Prosper Buranelli – Fletcher

RELOCATION
Decision to Relocate the Japanese Americans by Roger Daniels. – anti-US policy
America's Concentration Camps by Bosworth – enemy aliens
Betrayal from the East by Alan Hynd

SHIPS
Fighting Ships of WWII by Jane's
Battleships of 20th Century by Jane's
Hunter Killer by William T. Y'Blood. -- CVEs
The Big E - The Story of the USS Enterprise by Cdr Edward Stafford.
The Carrier War by Clark G. Reynolds (Time-Life)
Saratoga, CV-3; Illustrated History of the Legendary Carrier by John Fry
That Gallant Ship, USS Yorktown CV-5 by Robert J. Cressman
The Ship that Held the Line -- The USS Hornet and the First Year of the Pacific War by Lisle A. Rose
Queen of the Flat-Tops by Stanley Johnston -- USS Lexington and Coral Sea
The Fightin'est Ship: The story of the cruiser Helena by C. G Morris
Task Force 57 by Peter Smith -- British Pacific Fleet, 1945
Hunter Killer by William Y'Blood – CVEs, Atlantic
The American Battleship by S L Morison with Norman Polmar
*The World Encyclopedia of Battleship*s by Peter Hore
WWII Warship Guide by Robert Hewson
Warships by H P Willmott
Shinano! Sinking of Japan's Supership by Joseph Enright

STRATEGY
Planning for Victory World War II by Monro MacCloskey
Achieving Victory World War II by Monro MacCloskey
World War II Battle Plans edited by Stephen Badsey
The Nimitz Gray Book, War Diary maintained CINCPAC staff

SUBS
Silversides by Robert Trumbull
Silent Victory - Submarine War Against Japan by Clay Blair - 2 vols
Bowfin by Edwin P. Hoyt.
1941-43: Against The Odds by W.J. Holmes.
Fleets of World War II by Richard Worth.

MISCELLANEOUS
Japanese War Machine edited by S.L. Mayer.
Illustrated Weapons & Warfare - 24 volumes
The Bluejackets' Manual by U.S. Naval Institute. – WWII edition
World War II Desk Reference by Eisenhower Center -- better for ETO
Japanese Prisoners of War -- scholarly
Flag of Our Fathers by James Bradley and Ron Powers -- Saribuchi Flag Raising
Torpedo Junction by Homer H. Hickam,Jr -- USCG vs. U-boats, East Coast, Jan-Aug'42
War in the Pacific by Jerome T. Hagen, 2 vol -- articles
The Technology of World War II by Sean Sheehan -- shallow
Goodbye, Darkness: Memoir of the Pacific War by William Manchester. -- personal
Tigers of the Sea: 7 Vol, video
Remember Pearl Harbor: 5 Vol, video
Victory at Sea, by Richard Rogers with CBS Symphony Orchestra. -- inspiration.
The World Almanac And Book of Facts For 1943 pub New York World Telegram
Weapons That Wait -- Mine Warfare in the U.S. Navy by Gregory Kemeny Hartmann -- mine warfare including last 4 months WWII Pacific
V-J Day, Time 60th Anniversary Tribute
Suicide Squads by Richard O'Neill – midget subs plus
Suicide Submarine: the Kaiten Weapon by Yutaka Yokota with J.D. Harrington
The Secret War by Francis Russell, Time-Life Books
Prelude to War by Robert Elson, Time-Life Books
Fatal Victories by William Weir
The Good War: Readings by James Olson and Randy Roberts
Japan's War by Edwin Hoyt
Radioman by Carol Hipperson

I don't have these that are recommended by readers
"Winning the War in the Pacific" by James J. Tritten
"The Fleet the Gods Forgot" - Asiatic Fleet
"The Hands of Fate" – PBYs of PatWing 10

Reference 261

Recent or Skimmed. These are sitting in the to-be-read shelf. I have skimmed through them and have occasionally used for research but cannot claim to have fully absorbed them.

"On the Treadmill to Pearl Harbor; Memoirs of Adm Richardson", told to Dyer
"Soldier and Statesman: George C. Marshall' by Harold Faber
"On the Spot Reporting; Radio Records History" by Gordon & Falk
"Clear the Bridge" by Richard O'Kane - *Tang*
"Pacific Destiny" by Richard O'Connor - 1776-1941
"Touched with Fire : The Land War in the South Pacific" by Eric M. Bergerud
"War Without Mercy : Race & Power in the Pacific War" by John Dower
"American Military History, Oxford Companion" edited by John Chambers
"Japan's Imperial Conspiracy" by David Bergamini
"The Second World War: Vol I, II" by Winston Churchill and Editors of Life
"Flyboys" by James Bradley
"The Second World War" by John Keegan
"Our Mother's War" by Emily Yellin
"In Defense of Internment" by Michelle Malkin
"Illustrated World War II Encyclopedia" by Eddy Bauer
"The Greatest Generation Speaks" by Tom Brokaw
"Defining Moments: World War II" by Alex Hook
"Stillwell and the American Experience in China 1911-45" by Barbara Tuchman
"The Forrestal Diaries" – edit by Walter Millis
"Churchill, A Biography" by Roy Jenkins
"Japan's War, The Great Pacific Conflict" by Edwin P. Hoyt
"No Right to Win" by Ronald W. Russell – Midway
"Shattered Sword" by J. Parshall and A Tulley – Midway
"The Pacific War" by William B, Hopkins – comprehensive
"Dawn Like Thunder" by Robert J. Mrazek – Torpedo Squadron Eight
"The Pacific War" by William B. Hopkins -
"Pacific Crucible" by Ian W Toll – Pacific, 1941-42

Warning : these books contain gross inaccuracies.
[*This list is not included, but be aware there are errors out there.*]

Note that very little is presented on the website, hence in this book, without some authority's documentation, The only claim is to have put this in a form that I wish were available 60 years ago when I starting to read every book that could be found about WWII.

Readers are a source of information with sometimes unpublished facts, insights and corrections. Many a web page has been created as the result of an interesting query from a reader.

Pictures are mostly taken from the internet because they can be of better quality than those scanned from a book.

About the Author

James Bauer grew up in Pennsylvania during the war while an uncle served in the Pacific to inspire a lifetime of study of the naval aspects of the early years of the Pacific War.

After service in the submarine and destroyer fleets, he graduated from Penn State U. and pursued a career in engineering systems for companies and as a consultant while attending seven colleges and universities and can list BS, PE, MA, CFPIM, CCP, CDM after his name.

On retiring to a farm in Marshalltown, Iowa, he discovered the boyhood home of Admiral Frank Jack Fletcher. Recognizing him as a key figure in his study of WWII in the Pacific and finding that even in his hometown the Admiral is little known, Bauer set up a website to tell of his accomplishments   http://www.ww2pacific.com

The website has expanded to display other lesser known aspects of that war that had been gleaned over the years and is told for a new generation that may know of the great victories, but does not know of the early years when we were losing. It was in those early years where Fletcher with skill, courage and more than a little luck was able to shorten that war and save thousands of lives. Fletcher deserves a book to spread the word and Bauer has done it.

Manorborn Press,
2490 248th Street
Marshalltown, Iowa
Copyright 2018

www.ingramcontent.com/pod-product-compliance
Lightning Source LLC
Chambersburg PA
CBHW061634040426
42446CB00010B/1409